Renewable Energy for the Arctic

This book explores various facets of the transition to renewable energy in the Arctic region. It critically examines the adverse effects of fossil fuel extraction and use, environmental and social impacts of climate change, and the possibility of a low carbon energy system through innovation and technology.

Drawing together a diverse range of contributors and considering a range of new energy sources, this volume also looks at the scale of the transition challenges in the Arctic energy production and use, the necessary flexibility to balance energy demand and supply, the need of a more integrated energy infrastructure, and the new energy business models, health and safety, and quality standards for the region. Finally, it examines the transit and influence between Arctic and non-Arctic countries, in terms of growth, partnerships and new dynamics of a transitioning process to a sustainable energy system.

Focusing on specific case studies that represent the most relevant energy projects in the region, this book will be of great interest to students and scholars of energy policy and transitions, climate change, global business and sustainable development.

Gisele M. Arruda is a principal researcher in circumpolar studies in the field of Energy, Arctic, Climate Change, Environment and Society. She lectures on and researches Arctic Energy Management Systems, Renewable Energy, Climate Change and Arctic Studies (Social Impacts of energy and resources development) and Education. She researches at Oxford Brookes University and she lectures at the Global Energy and Sustainability Programme at Coventry University, UK.

Routledge Explorations in Energy Studies

Energy Poverty and Vulnerability
A Global Perspective
Edited by Neil Simcock, Harriet Thomson, Saska Petrova and Stefan Bouzarovski

The Politics of Energy Security
Critical Security Studies, New Materialism and Governmentality
Johannes Kester

Renewable Energy for the Arctic
New Perspectives
Edited by Gisele M. Arruda

Renewable Energy for the Arctic

New Perspectives

Edited by Gisele M. Arruda

Routledge
Taylor & Francis Group

LONDON AND NEW YORK

First published 2019 by Routledge

2 Park Square, Milton Park, Abingdon, Oxfordshire OX14 4RN
52 Vanderbilt Avenue, New York, NY 10017

Routledge is an imprint of the Taylor & Francis Group, an informa business

First issued in paperback 2020

British Library Cataloguing-in-Publication Data
A catalogue record for this book is available from the British Library

Library of Congress Cataloging-in-Publication Data
A catalog record has been requested for this book

ISBN: 978-0-8153-8732-9 (hbk)
ISBN: 978-0-367-51173-9 (pbk)

Typeset in Times New Roman
by Wearset Ltd, Boldon, Tyne and Wear

Contents

Figures

Tables

Contributors

Tom Marsik is Associate Professor of Sustainable Energy at the University of Alaska Fairbanks, College of Rural and Community Development, Bristol Bay Campus and the official world record holder for the Tightest Residential Building. He is also a recipient of Alaska's Top Forty Under 40 award. He has an MS in Electrical Engineering from the Czech Technical University in Prague, and a PhD in Engineering from the University of Alaska Fairbanks. He has dedicated both his personal and professional lives to sustainable energy and related education. He has numerous publications and has served on energy-related committees, including the Consumer Energy subcommittee of Alaska's Governor Bill Walker's Transition Team. He has a loving wife and a four-year-old daughter whose face keeps reminding him to work hard on helping develop a sustainable future.

Nathan Wiltse is Policy Program Manager and Primary Economist at the Cold Climate Housing Research Center in Fairbanks Alaska. He has a BA in Economics from St. Olaf College and an MS in Mineral Economics from the University of Alaska Fairbanks. On behalf of the Cold Climate Housing Research Center he has worked with the Alaska Housing Finance Corporation and tribal entities on numerous issues related to energy efficiency and housing in Alaska.

Gisele M. Arruda has extensive experience as a Principal Researcher in circumpolar studies and in the field of energy, arctic, climate change, environment and society. She lectures and researches on arctic energy management systems, renewable energy, environmental management, arctic studies (social impacts of energy and resources development) and education. She belongs to the research team at Oxford Brookes University, to the Global Energy and Sustainability programme at Coventry University, she is a Fellow of Higher Education Academy, a Fellow of the Royal Geographical Society, and an alumnus (Master's degree and pHD) of Harvard University, Oxford University and Oxford Brookes University in Human Geography, Law, Management and Policy. She is the editor-in-chief of *Energy, Arctic & Climate Change International Journal,* peer-reviewer for other international journals and a PhD supervisor. orcid.org/0000-0002-4541-3592.

Feb M. Arruda is a Principal Researcher in circumpolar studies and Director of Anvivo Research (Anvivo.org). He has undertaken fieldwork in Norway, Canada, Greenland, Iceland, Europe, the US and Latin America. He is an ecologist with a MA in Ecology and Agriculture.

Julianna Mae Hogenson holds a MA in Sustainable Energy Science from Reykjavik University and Iceland School of Energy. Her BSc was obtained at the University of Lethbridge in Archaeology and Geography, Alberta, Canada. Throughout her academic career she has attended post-graduation programmes at Tsinghua University in Beijing, China and multiple field schools in the Middle East and North America. An internship at Icelandic GeoSurvey (ISOR) led to a thesis project focused upon geothermal surface research in Iceland. Her passion for learning and education inspired her contribution to this book.

Enzo A. J. Diependaal has a MSc in Sustainable Energy Engineering at the Iceland School of Energy, part of Reykjavik University. He broadened his degree through the nuclear engineering winter school at St. Petersburg Polytechnic and Arctic energy and petroleum sciences at the University Centre in Svalbard. His thesis and research at the University of Ontario Institute of Technology focuses on experimental nuclear technology analysis and priority setting for off-grid mining in the Canadian Arctic, using The Integrated MARKAL-EFOM System (TIMES) model generator. He holds a BSc in Aviation Engineering from the Amsterdam University of Applied Sciences.

Hans-Kristian Ringkjøb is a PhD candidate at the Geophysical Institute, University of Bergen. His research is in renewable energy, with a focus on modelling of energy systems using the TIMES modelling framework. He holds a double MA in renewable energy from École Polytechnique (France) and Instituto Superior Técnico (Portugal), achieved through the MSc RENE programme offered by InnoEnergy.

Sarah Sternbergh is a professional engineer working for Tetra Tech Canada Inc., with a Master's degree in Sustainable Energy Engineering from Reykjavik University in Iceland. She currently resides and works in Canada's Yukon Territory. She works on a diverse array of projects related to geological and energy engineering including energy projects such as low temperature geothermal exploration and development.

Dyveke F. Elset holds a MA in International Political Economy from University of Kent's Brussels School of International Studies and a BA in International Relations from University of Birmingham, UK. Her professional and academic field of interests are energy, climate change and Arctic issues. She wrote her Master thesis on the role of multinational corporations in the renewable energy transition and has previously published the article 'Bridging environmental and energy security interests in the Arctic' in *Energy, Arctic & Climate Change Journal*. She is currently based in Oslo, Norway, working as a Communication Adviser for Nordic Innovation.

Melania Milecka-Forrest leads the Strategy and Operations Management teaching team at Coventry University London. Skilled in the complex Lean Six Sigma business improvement methodology and Knowledge Management theories, she has used quality and process management tools to successfully redesign corporative strategies and MBA courses. She has worked closely with major awarding bodies and publishers in the UK to embed professional qualification structures, assessment strategies and British academic standards. She believes in innovation and cutting-edge technologies to engage modern students and professionals in high performance and quality management programmes relevant to the future of Arctic energy industry.

Natalia Rocha-Lawton is Lecturer in HR and Organisational Behaviour at the Faculty of Business and Law, Coventry University. She holds a PhD in Sociology from the University of Warwick. Prior to joining academia, she worked at the Secretariat of Energy for the Mexican Government. Her research focuses on the social implications of economic deregulation of the Energy Sector on employment relations, international human resources management, gender and diversity. She has been guest editor and author for *Work Organisation, Labour & Globalisation* journal and an author for SAGE and Cambridge University Press. orcid.org/0000–0002–0580–3457.

Nicholas Craig is a fellow of the Polar Research and Policy Initiative (PRPI), specialising in energy and climate change. Originally from the UK and now based in Denmark, he contributes across PRPI's portfolio of work, with a focus on sustainable development in the Arctic and the UK's energy relationship with Arctic States. He is also currently completing a postgraduate degree in Climate Change at the University of Copenhagen and previously studied at the University of Helsinki and the University of Exeter. He has undertaken a range of climate change fieldwork in Greenland, Finland, the French Alps and Brazil. He has a background in renewables and energy policy, and previously worked on the Britain Stronger in Europe campaign. In Copenhagen he also works for Sustainia, a sustainability think tank and advisory. He is particularly fascinated by the new opportunities that a changing Arctic brings and the interconnectivity of natural, commercial and social factors in the region.

Part I

Overview

Low carbon Arctic energy system

1 A low carbon Arctic energy system?

Challenges, opportunities, and trends

Tom Marsik and Nathan Wiltse

Introduction

The Arctic has a unique position in the global energy system. Not only does it contribute to the global fossil fuel use through its own consumption and exports, it also greatly suffers the consequences of global fossil fuel use due to climate change and polar amplification (NASA, 2013). The warming of the Arctic is posing numerous challenges for inhabitants, including the need to sometimes relocate whole communities (Mooney, 2015). Impacts of climate change coupled with other issues, such as depleting finite resources, has led Arctic stakeholders to adopt initiatives for decreasing fossil fuel use and reducing carbon emissions. This chapter, as a starting point of the discussion in the whole book, provides a brief overview of the current energy situation as well as challenges and opportunities for achieving a low carbon Arctic energy system.

Energy use in the Arctic

Current situation

Arctic energy consumption

Excluding Russia,[1] annual energy consumption from all energy types in the Arctic is approximately equivalent to 129,000 KTOE (thousand tonnes of oil equivalent) or about 12 percent of the consumption of the European Union's 28 countries (see Figure 1.1). One KTOE is equivalent to 41,868 gigajoules. A majority (71 percent) of the energy consumption occurring in the Arctic is in European nations.[2] Alaska (US) consumes about two-thirds the energy that the combined Canadian areas of British Columbia, Yukon Territory, Northwest Territory, and Nunavut (British Columbia (BC) & Territories)[3] consume.

Consumption per capita

A more useful metric for comparing the adoption of energy efficiency by Arctic regions is annual energy consumption per capita. Looking at these values,

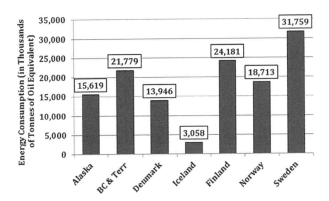

Figure 1.1 Annual energy consumption by Arctic region (KTOE), 2015.

Sources: Comprehensive Energy Use Database (NRCan, 2017); Eurostat (Eurostat, 2018); State Energy Data System (EIA, 2017).

Alaska and Iceland are the main per capita energy consumers (about 21 and nine tonnes of oil equivalent per person respectively), while Scandinavian nations range from 2.5 to 4.4 tonnes of oil equivalent (TOE) per person (see Figure 1.2). Alaska's low population, and large petroleum and mining industries mean that over half of its per capita energy consumption comes from the industry sector. Further, Alaska has an abundance of cheap (subsidized) natural gas used by the two main population centers that account for approximately half of its population.

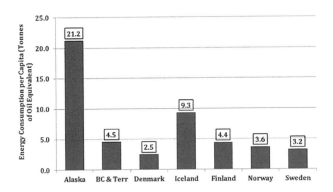

Figure 1.2 Annual energy consumption per capita by Arctic region (TOE), 2015.

Sources: Comprehensive Energy Use Database (NRCan, 2017); Eurostat (Eurostat, 2018); State Energy Data System (EIA, 2017); Alaska population estimates (DOL, 2018); Canadian population estimates (CANSIM, 2017); Denmark population estimates (Statistics Denmark, 2018); Finland population estimates (Statistics Finland, 2017); Iceland population estimates (Statistics Iceland, 2017); Norway population estimates (Statistics Norway, 2017); Sweden population estimates (Statistics Sweden, 2017).

Approximately half of Alaska's per capita consumption is natural gas. Iceland has an abundance of cheap geothermal energy that drives the electrical production and derived heat.

Consumption by energy type

While fossil fuels dominate energy consumption in the circumpolar region as a whole, the individual Arctic regions have differing access to energy resources and use a variety of fuel types (see Table 1.1). The least fossil fuel dependent region is Iceland, where about three-fourths of the energy consumption come from renewable sources. In contrast, almost all of Alaska's energy consumption is covered by fossil fuels. Overall, Arctic regions meet about 44 percent of their consumption with renewable energy.[4]

Consumption by sector

For the majority of Arctic regions, industrial energy consumption is the major consumer of energy (see Table 1.2). With the exception of Iceland, the second major end use sector is transportation. For all Arctic regions these two end use sectors comprise between 50 percent and 82 percent of energy consumption. While the energy consumption of aviation often receives press, for all five European regions, road transportation energy consumption exceeds both international and domestic aviation combined.

Trends in energy use in the Arctic

Trends in consumption

Over the 25-year period from 1990 to 2015, most Arctic regions saw a period of increased energy consumption, with the overall energy consumption for the seven Arctic regions (excluding Russia) increasing by 8,332 KTOE for the period. Some regions saw early increases in overall energy consumptions followed by decreases. These peaks occurred at different times for different regions. Sweden's, for example, came in the mid-1990s while Alaska's came in the mid-2000s. Iceland's energy consumption has more than doubled in the past 25-years on a trajectory of steady growth (see Table 1.3).

Trends in consumption per capita

In observing energy consumption over time, a per capita examination can be helpful. Iceland's doubling of energy consumption is actually mirrored in its per capita data, indicating that the energy consumption is not due to an increase in population, but rather primarily due to growth in the industrial sector (see Table 1.4). On the other hand, BC & Territories, which showed fairly stable energy use in Table 1.3, show a significant decrease (about 33 percent) in energy consumption

Table 1.1 Annual energy consumption by fuel type for Arctic regions (KTOE), 2015

Energy type	Alaska	BC & Terr.	Denmark	Iceland	Finland	Norway	Sweden
Renewable energies	484 3%	8,194 38%	4,081 21%	2,196 72%	11,968 49%	10,119 54%	19,789 62%
Derived heat[1] (non-renewable)	0 0%	0 0%	1,382 10%	0 0%	2,127 9%	243 1%	1,066 3%
Electrical energy (non-renewable)	1,142 7%	404 2%	1,041 7%	0 0%	1,875 8%	549 3%	602 2%
Natural gas	7,734 50%	3,167 15%	1,467 11%	0 0%	843 3%	482 3%	750 2%
Solid fossil fuels	214 1%	180 1%	118 1%	94 3%	603 2%	610 3%	1,036 3%
Total petroleum products[2]	6,048 39%	9,834 45%	5,837 42%	766 25%	6,716 28%	6,700 36%	8,515 27%
Waste[3] (non-renewable)	0 0%	0 0%	21 0.1%	0 0%	48 0.2%	10 0.1%	0 0%

Sources: Comprehensive Energy Use Database (NRCan, 2017); Eurostat (Eurostat, 2018); State Energy Data System (EIA, 2017); Statistics Canada (CANSIM, 2017).

Notes

1 The authors have separated renewable energy sourced derived heat and added it to the renewable energy numbers. The definition of derived heat from the Eurostat dataset is the following: "Derived heat covers the total heat production in heating plants and in combined heat and power plants. It includes the heat used by the auxiliaries of the installation which use hot fluid (space heating, liquid fuel heating, etc.) And losses in the installation/network heat exchanges. For auto-producing entities (= entities generating electricity and/or heat wholly or partially for their own use as an activity which supports their primary activity) the heat used by the undertaking for its own processes is not included." (Eurostat, 2017).

2 Total petroleum products excluding natural gas and solid fossil fuels.

3 The definition of Waste (non-renewable) from the Eurostat dataset is the following: "Waste (non-renewable) consists of materials coming from combustible industrial, institutional, hospital and household wastes such as rubber, plastics, waste fossil oils and other similar types of wastes, which can be either solid or liquid." (Eurostat, 2017).

Table 1.2 Annual energy consumption by end use sector for Arctic regions (KTOE), 2015

Sector	Alaska	BC & Terr.	Denmark	Iceland	Finland	Norway	Sweden
Agriculture	0	597	634	39	689	304	350
	0%	3%	5%	1%	3%	2%	1%
Industry	8,387	6,951	2,110	1,461	10,698	5,909	11,528
	54%	32%	15%	48%	44%	32%	36%
Residential	1,237	3,597	4,254	392	4,898	3,881	7,197
	8%	17%	31%	13%	20%	21%	23%
Transportation	4,322	8,370	4,949	521	4,791	5,470	8,668
	28%	38%	35%	17%	20%	29%	27%
Commercial and other sectors	1,673	2,263	1,998	645	3,104	3,149	4,016
	11%	10%	14%	21%	13%	17%	13%

Sources: Comprehensive Energy Use Database (NRCan, 2017); Eurostat (Eurostat, 2018); State Energy Data System (EIA, 2017).

Table 1.3 Historical annual energy consumption (KTOE), 1990 to 2015, by Arctic region

Year	Alaska	BC & Terr.	Denmark	Iceland	Finland	Norway	Sweden	Total
1990	14,719	22,235	13,453	1,410	21,650	16,088	31,160	120,715
1995	17,746	24,996	14,818	1,490	21,974	16,946	35,051	133,020
2000	18,703	26,865	14,718	1,871	24,316	18,094	34,973	139,540
2005	20,102	26,701	15,499	2,009	25,185	18,580	33,659	141,734
2010	16,461	23,342	15,519	2,644	26,247	19,602	34,077	137,890
2015	15,619	21,779	13,946	3,058	24,181	18,713	31,759	129,053

Sources: Comprehensive Energy Use Database (NRCan, 2017); Eurostat (Eurostat, 2018); State Energy Data System (EIA, 2017).

Table 1.4 Historical annual energy consumption per capita (TOE), 1990 to 2015, by Arctic region

Year	Alaska	BC & Terr.	Denmark	Iceland	Finland	Norway	Sweden
1990	26.6	6.6	2.6	5.6	4.3	3.8	3.6
1995	29.5	6.5	2.8	5.6	4.3	3.9	4.0
2000	29.8	6.5	2.8	6.7	4.7	4.0	3.9
2005	30.1	6.2	2.9	6.8	4.8	4.0	3.7
2010	23.1	5.1	2.8	8.3	4.9	4.0	3.6
2015	21.2	4.5	2.5	9.3	4.4	3.6	3.2

Sources: Comprehensive Energy Use Database (NRCan, 2017); Eurostat (Eurostat, 2018); State Energy Data System (EIA, 2017); Alaska population estimates (DOL, 2018); Canadian population estimates (CANSIM, 2017); Denmark population estimates (Statistics Denmark, 2018); Finland population estimates (Statistics Finland, 2017); Iceland population estimates (Statistics Iceland, 2017); Norway population estimates (Statistics Norway, 2017); Sweden population estimates (Statistics Sweden, 2017).

per capita over the 25-year period. This was driven primarily by decreases in overall industrial energy consumption. Alaska is notable as its energy consumption per capita is between four and five times that of BC & Territories. Starting in 2005, Alaska saw a 30 percent decrease in per capita energy consumption over the following ten years ending in 2015. This decrease was primarily driven by nearly equal decreases in industrial energy consumption and transportation consumption over that period. However, with an energy consumption per capita of 21.2 TOE in 2015, Alaska residents still consume nearly ten times the energy of Danish residents and more the twice the energy of Iceland residents. As mentioned previously, over half of Alaska's per capita consumption comes from the industrial sector.

Trends in consumption by energy type

As noted previously, overall the energy consumption for the seven Arctic regions (excluding Russia) has increased over the past 25 years. The consumption of renewable energy has increased by 16,025 KTOE, while the fossil fuel consumption saw a decrease of 7,692 KTOE over the 25-year period (see Table 1.5).

Table 1.5 Historical annual energy consumption (KTOE), 1990 to 2015, by fuel type

Year	Renewable energies	Derived heat (non-renewable)	Electrical energy (non-renewable)	Natural gas	Solid fossil fuels	Total petroleum products	Waste (non-renewable)
1990	40,809	4,388	6,740	15,471	4,577	48,724	15
1995	46,159	5,979	6,558	19,982	4,337	49,984	16
2000	49,917	6,060	7,419	20,378	4,118	51,597	43
2005	51,606	6,228	7,768	19,450	3,788	52,810	84
2010	53,272	7,280	8,537	16,135	3,315	49,277	80
2015	56,833	4,818	5,611	14,443	2,855	44,416	79

Sources: NRCan—Natural Resources Canada (NRCan, 2017); Eurostat (Eurostat, 2018); State Energy Data System (EIA, 2017); Statistics Canada (CANSIM, 2017).

Table 1.6 Changes in renewable and non-renewable annual energy consumption (KTOE), 1990 to 2015

Energy type	Alaska	BC & Terr.	Denmark	Iceland	Finland	Norway	Sweden
Renewable energies	23 +5%	357 +5%	2,851 +232%	1,460 +198%	4,851 +68%	893 +10%	5,591 +39%
Non-renewable energies	872 +6%	−813 −6%	−2,359 −19%	188 +28%	−2,321 −16%	1,732 +25%	−4,990 −29%

Sources: Comprehensive Energy Use Database (NRCan, 2017); Eurostat (Eurostat, 2018); State Energy Data System (EIA, 2017).

Increases in renewable energy consumption were driven by the countries of Denmark, Iceland, Finland, and Sweden (see Table 1.6). Denmark's consumption of renewable energy grew from 1,230 KTOE in 1990 to 4,081 KTOE in 2015, a 232 percent increase. Iceland's consumption of renewable energy grew from 738 KTOE in 1990 to 2,198 KTOE in 2015, a 198 percent increase. Finland and Sweden saw increases of 4,851 KTOE and 5,591 KTOE respectively over the 25-year period. Increases for the four regions were larger than changes in population, resulting in an increase in per capita renewable energy use.

Trends in consumption by sector

The overall trends for the consumption by sector in the Arctic are shown in Table 1.7. The agricultural sector saw declines in energy consumption from all Arctic regions except BC & Territories and Iceland. The industrial sector saw a decrease in energy consumption driven primarily by BC & Territories, Denmark, and Sweden. The commercial and other sector saw growth in energy consumption driven primarily by Finland and Norway.

The residential sector saw an increase in energy consumption driven by all Arctic regions except Finland. Finland saw a decrease in residential energy consumption over the period from 1990 to 2015. The largest decreases in per capita energy consumption in the residential sector over the period were seen in BC & Territories at 27 percent, Alaska at 22 percent, Finland at 17 percent, and Norway at 12 percent. This is indicative of energy efficiency efforts in the residential sector for those regions.

The largest growth in energy consumption over the period from 1990 to 2015 was found in the transportation sector. Energy consumption for the transportation sector grew 22 percent over the 25-year period. While all Arctic regions contributed to this increase, the primary drivers were BC & Territories, Norway, and Sweden. On a per capita basis, both Alaska and BC & Territories saw decreases of 24 percent and 9 percent respectively. The largest per capita increases were seen in Iceland and Norway (39 percent and 17 percent respectively).

Table 1.7 Historical annual energy consumption (KTOE), 1990 to 2015, by sector

Year	Agriculture	Industrial	Residential	Transportation	Commercial and other sectors
1990	3,250	48,479	24,405	30,522	14,059
1995	3,233	54,225	26,738	33,033	15,790
2000	2,930	58,619	25,232	35,828	16,931
2005	2,832	56,660	26,291	38,854	17,097
2010	2,982	49,743	28,530	38,322	18,313
2015	2,613	47,045	25,457	37,091	16,848

Sources: Comprehensive Energy Use Database (NRCan, 2017); Eurostat (Eurostat, 2018); State Energy Data System (EIA, 2017).

Challenges

Cold climate

The severity of the Arctic brings several challenges. The main one is that a significant amount of energy needs to be used for heating buildings (CCHRC, 2017). Further, transportation requires extra energy expenditure due to additional losses (for example the higher rolling resistance of tires in the cold) and extra measures (for example, plugging in engine block heaters to warm up combustion engines via electric resistive heating before they are started).

Long distances

The Arctic has a large number of remote communities with significant distances between them (Hossain et al., 2016), which bring several challenges. Transportation of people often has to be done by air, which is energy intensive. Extra energy has to be expended to transport food and other supplies to these communities. Fossil fuels also have to be transported to these communities, which further increases the carbon footprint of these energy sources because additional fuel has to be used for their transportation. The combination of long distances and cold climate further increases energy use, as oceans and rivers, often used as supply routes during warmer parts of the year, freeze up and more energy-intensive means of transportation have to be used during the colder parts of the year.

Small communities

Many Arctic communities are small in size. Combined with their remoteness, this prevents many projects that rely on economies of scale. For example, the expense of building roads and transmission lines between communities often cannot be justified. The lack of roads leads to more energy-intensive means of transportation and the lack of transmission lines leads to isolated microgrids.

Small grids cannot take advantage of economies of scale of some renewable energy technologies as discussed later in this chapter. The lack of transmission lines also prevents communities from accessing renewable energy resources that might be available elsewhere. The small size of communities combined with their remoteness also poses challenges with regard to accessing skilled workforce to install and maintain complex energy systems.

Besides increased use of fossil fuels, these challenges also pose other issues, such as decreased reliability, energy insecurity, and increased energy costs. However, these challenges also present some opportunities as described in the following section.

Opportunities

Energy efficiency

For the purposes of this chapter, "energy efficiency" refers to any measures resulting in decreased use of energy. This includes both the use of technology (e.g., more efficient lighting and appliances) and changes to human behavior (e.g., turning off lights when not needed).

Energy efficiency presents a vast opportunity for the Arctic (The Energy Efficiency Partnership, 2013; Arctic Energy Alliance, 2017). Our predecessors in the Arctic, before the era of inexpensive fossil fuels, demonstrated that one can live with significantly less energy than we use today (Lee and Reinhardt, 2003). While energy can increase quality of life when supporting basic needs, this correlation doesn't necessarily apply to the levels of energy we use in the modern world. For example, many European countries rank higher in happiness than the United States (Helliwell et al., 2013) despite—or perhaps because of—the fact that their energy use per capita is significantly lower than in the US (Ristinen and Kraushaar, 2006). These statistics suggest that energy use can be reduced while maintaining, or even improving, quality of life.

A vast range of energy efficiency technologies and approaches applicable to the Arctic exists. Some examples include: upgrading to more efficient lighting, removing light bulbs in overly lit areas, use of occupancy sensors, use of programmable thermostats, adjusting ventilation schedules, utilizing heat recovery ventilation, air sealing building envelopes, insulating, replacing old appliances with new efficient ones, upgrading to more efficient vehicles, using by-product heat from power plants, upgrading motors in industrial processes, optimizing airplane routes and schedules, and many more. Those approaches that rely on behavioral changes, such as riding a bicycle instead of driving a car, often include additional benefits, such as health, besides saving energy.

Economics

Economics of different energy efficiency approaches can vary widely, but many cost-effective improvement options exist. For example, one can look at the economics of replacing a 60 watt (W) incandescent light bulb with a 10 W light emitting diode (LED) lamp in Kotzebue (an Arctic community in Alaska). Upgrading from a 60 W incandescent light bulb to a 10 W LED represents a savings of 50 W. Assuming the light bulb is on six hours per day, this upgrade saves 0.3 kWh per day, or about 110 kWh in a year. The fuel cost in Kotzebue is about US$0.20/kWh,[5] so the annual savings are about US$22 per bulb used in Kotzebue. A 10 W LED can be bought for about US$3, including shipping. Labor is neglected because the amount of time involved in replacing a standard screw-in light bulb is minimal, plus some people choose to upgrade when an incandescent light bulb burns out and needs to be replaced anyway. With a US$3 initial investment and US$22 annual savings, one is looking at the first year return of about 700 percent, which is extraordinary.

Examples of use

Many energy efficiency success stories exist in the Arctic. Finland ranks among top countries thanks to energy savings achieved by cogeneration of heat and electricity, voluntary energy efficiency agreements, and systematic energy auditing (MEAE, 2017). Sweden has implemented a successful "Programme for Energy Efficiency in Energy Intensive Industry (PFE)" program with significant energy savings (Swedish Energy Agency, 2017). Russia has ongoing energy efficiency efforts in its industry (Mokveld, 2011). And Alaska's Home Energy Rebate and Weatherization programs have saved households about 33 percent of annual energy consumption (Goldsmith et al., 2012).

Renewable energy

Historically, renewable energy in the Arctic has been used on a small scale, such as wood for space heating, seal oil for lighting, and sail boats for transportation. While renewable energy resources are considered abundant in the Arctic, the feasibility of harvesting these resources on a larger scale is a different issue. The renewable energy sources considered in this chapter are river hydro, wind, solar, biomass, geothermal, tidal, and ocean waves. There are many factors that need to be considered when evaluating renewable energy sources for the Arctic. Important factors to consider include: magnitude of resource, proximity of the resource to the point of use or existing distribution lines, land ownership and public use issues, permitting, maturity of available technologies, suitability of technologies for cold temperatures, aesthetics, longevity of technologies, skills of local workforce to operate and maintain installed systems, life-cycle economics and environmental impacts, effects on grid stability, risks of catastrophic failures and other safety issues, acceptance of the project by local culture, and other factors. The following sections cover basic aspects of specific renewable energy resources in the Arctic. The emphasis is on Arctic-specific considerations, as existing publications cover general information about these renewable sources (Boyle, 2004; Sorensen, 2004).

River hydro

The Arctic has lakes and rivers that could support numerous hydroelectric pro-jects of large capacity, but many are located in sensitive ecosystems or very far from population centers where the energy could be used. Another problem with hydropower systems in the Arctic is the formation of ice and the addi-tional measures needed to prevent blockage or damage to the mechanical com-ponents. These measures often lead to reduced power production (Gebre et al., 2013). Unfortunately, this decrease in output occurs in winter when the power demand is typically the highest. Despite the challenges, some Arctic regions have taken a strong advantage of the available resource. In Norway, hydro-power accounts for nearly 99 percent of the total electricity production (Gebre et al., 2013).

While the emphasis of this section has been on regular hydro systems, which have a dam (or in some situations a natural feature that serves as a dam), another type of river hydro to mention is hydrokinetic systems, which have no dam and consist of one or more in-stream turbines placed in the natural water current. While hydrokinetic technology does not exist yet in a commercial, reliable form for Arctic conditions, it is an emerging technology with a significant research interest. Debris management (including ice) is one of the issues that researchers are working on (ACEP, 2017).

Wind

Despite cold climate challenges, such as icing of blades, wind energy presents a significant opportunity for the Arctic. Figure 1.3 maps the wind resource for the Arctic and surrounding areas. As seen in the figure, overall, the Arctic has a sub-stantial wind resource.

On a smaller scale, there are significant variations in the wind resource. There are many areas with an excellent resource, however, the issue is that these areas are sometimes very distant from the potential point of use and existing electrical grids, or on top of steep terrain that creates logistical and cost barriers for successful projects. Further, because many communities in the Arctic are served by isolated microgrids, it can be difficult to take advantage of economies of scale. Economies of scale for wind projects have a significant effect for two main reasons: (1) in terms of the monetary amount per one watt of installed capacity, larger wind turbines are in general cheaper than smaller wind turbines; (2) larger wind turbines are typically taller where wind speed is higher. Therefore, from an economic standpoint, successful projects are typic-ally those of larger scale that are supplying electricity into larger grids. Cur-rently, Norway is building a 1 GW wind project called Fosen Vind, which is considered the largest one in Europe and the fourth largest in the world (Robarts, 2016). The construction is expected to be completed in 2020. Despite many positive aspects of the project, it is facing opposition from reindeer herders in the region (Losnes, 2016).

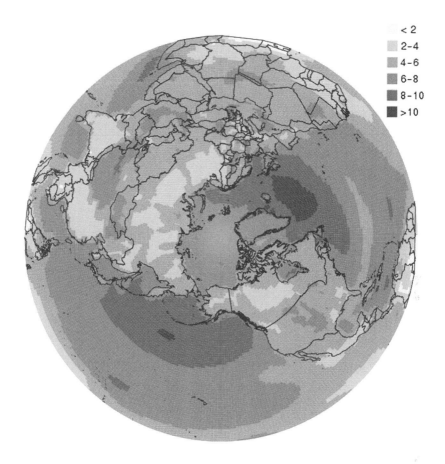

< 2
2-4
4-6
6-8
8-10
>10

Figure 1.3 Annual average wind speed in m/s at 50 m per NASA.
Source: Arctic Renewable Energy Atlas (NREL, 2017a).

Solar

Figure 1.4 maps the solar resource for the Arctic and surrounding areas. Despite a relatively poor resource, some opportunities exist in the Arctic to utilize solar energy through passive solar thermal, active solar thermal, or solar photovoltaic technologies.

"Passive solar" refers typically to the solar energy entering a building through windows. Since most buildings have windows anyway (for daylighting, etc.), passive solar can be an economical renewable energy source. Suitable orientation of a building and suitable distribution of windows might represent no added construction cost and increased passive solar gain, thus decreasing heating demand from other sources. However, adding windows specifically for the

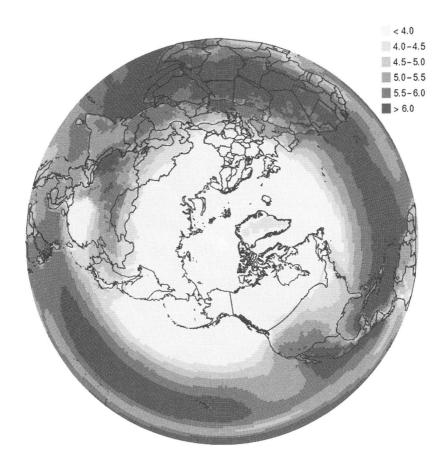

Figure 1.4 Annual average solar radiation in kWh/m^2/day at latitude tilt per NASA.
Source: Arctic Renewable Energy Atlas (NREL, 2017a).

purpose of the passive solar gain that are not needed for other purposes is not always advantageous in the Arctic because the additional heat loss through the windows can outweigh the passive solar gain.

"Active solar thermal" refers to the use of solar thermal panels and an active system (usually a pump) to transfer heat from the panels into a building through a moving fluid. Using an active solar thermal system for space heating in the Arctic is challenging because the resource is needed the most in winter when it is available the least. More often it is considered for hot water needs because hot water is needed all year round (including seasons when the solar resource is abundant). However, due to the combination of a relatively poor solar resource, colder ambient temperatures (which reduce the perform-ance of solar thermal panels) and other factors, the application of active solar thermal systems (even those used for hot water) in the Arctic has been limited (ACEP, 2013).

"Solar photovoltaic" (PV) refers to the technology for the direct conversion of solar energy into electrical energy. Despite some advantages in the Arctic, such as higher efficiency of PV panels in colder temperatures or additional radiation thanks to the reflection from snow, the disadvantage of the relatively poor solar resource typically far outweighs the advantages. However, while the levelized cost of energy[6] for past PV installations in the Arctic has been relatively high (ACEP, 2016), new opportunities are arising with the downward trend in costs of PV modules, especially in some remote communities with high energy costs (Roberts, 2017).

Biomass

While the availability of the resource varies widely throughout the Arctic and the location is sometimes very distant from population centers where the resource can be used, significant opportunities exist for utilizing plant material, animal material, or municipal waste for energy where available.

Plant biomass material used for energy is primarily in the form of trees and shrubs. While usable for space heating or electricity generation, producing electricity from wood on a small scale is challenging and is done mostly on a larger scale, especially where forestry residues are available from industrial processes. One of the leading countries in forest bioenergy is Finland, where about 25 percent of total energy consumption is produced with wood fuels (Koponen et al., 2015). While biomass is renewable, for it to be a sustainable resource it needs to be managed properly, as removing it faster than it is being replenished would eventually deplete the resource. In the Arctic, forest fire management practices can also provide an opportunity for harvesting biomass energy.

Animal material used for energy in the Arctic is primarily byproducts from fish processing plants. Oil extracted from the fish waste can be used for thermal processes (often those in the fish processing plants themselves) or sometimes for electricity production in diesel generators (AEA and REAP, 2016).

Municipal waste can be used for energy production, but the amount of the resource is very limited and can only supply a very small fraction of the needed energy (Marsik, 2009). Also, in remote Arctic communities, converting general municipal waste into useful energy is very challenging due to technological limitations on a small scale (Marsik, 2018), though some limited opportunities still exist for utilizing special types of waste, such as waste vegetable oil. Despite municipal waste being a limited resource, utilizing it when possible is a good practice as it helps with environmental issues related to landfills. In Sweden, more than 99 percent of all household waste is recycled or used to produce energy (Freden, 2017).

Geothermal

Most of the Arctic does not have any known shallow geothermal resource. The Arctic areas that do have a known shallow geothermal resource are often too distant from main population centers to justify resource development. However, some Arctic areas do have a viable geothermal resource of a significant magnitude and a primary example of such an area is Iceland.

Iceland has a unique location over a rift in continental plates and has a high concentration of volcanoes. This rich geothermal resource is being utilized for electricity as well as heating applications, which include space heating, domestic hot water, swimming pools, greenhouses, and others. About 65 percent of Iceland's primary energy is supplied from geothermal sources. Iceland's geothermal energy together with hydro, which provides about 20 percent of Iceland's primary energy, represents the world's highest share of renewable energy in any national total budget (Askja Energy, 2016).

Tidal

The Arctic has tidal resources comparable to the rest of the world. The two main ways to harvest tidal energy are barrages and hydrokinetic devices. Barrages are dams built across suitable estuaries and use basically the same technology as conventional river hydro systems harvesting the potential energy associated with different levels of water on either side of the dam. One of the significant considerations of such dams is the enormous impact on the marine ecosystems. Only one barrage exists in the Arctic and it is the Kislaya Guba experimental project in northwest Russia (Bernshtein, 1972; Myles, 2017).

Hydrokinetic devices are in-stream turbines harvesting the kinetic energy of tidal currents. Because no dam is needed, their impact on the ecosystem is considered to be significantly lower than barrages. Hydrokinetic devices are an emerging technology, with the world's first commercial-scale tidal generator, a 1.2 MW underwater turbine, operating in Northern Ireland from 2008 to 2016 (McDonnell, 2016). No hydrokinetic tidal systems currently exist in the Arctic, but there is a potential for future deployments with advancements in the technology and the understanding of environmental impacts.

Ocean waves

The Arctic, together with the Antarctic and tropical regions, has a relatively poor wave energy resource compared to the temperate zones (Gunn and Stock-Williams, 2012). Various wave energy converter technologies exist, ranging from generators powered by buoys oscillating vertically in waves to overtopping devices, which have reservoirs filled by incoming waves and the water then goes through a turbine back to the ocean surface (Myles, 2017). However, wave energy converters are an emerging technology, with the world's first wave farm opening in Portugal in 2008 and closing the same year (Kanellos, 2009). No functional wave farm currently exists in the Arctic, but Sweden is in the process of developing a wave energy park at Sotenäs (TidalEnergyToday, 2017).

Path toward low carbon Arctic

While a host of factors is used when making decisions regarding what approaches to use to reduce fossil fuel consumption (and thus reduce carbon emissions), economics is one of the most important ones. As explained in the Energy Efficiency section earlier, many energy efficiency improvement options exist that are very cost-effective in the current situation, some of them yielding annual returns in the order of hundreds of percent. On the other hand, renewable energy installations typically yield annual returns in the order of ones of percent (ACEP, 2016), or tens of percent in exceptional cases (AEA and REAP, 2016).

While energy efficiency is an extremely cost-effective resource when considering a wasteful situation as a starting point, it should be clarified that its cost-effectiveness decreases as the efficiency increases (for example, upgrading from double-pane windows to triple-pane is less cost-effective than upgrading from single-pane windows to double-pane). And energy efficiency cannot ever bring our energy use down to zero (e.g., it does not matter how efficient a vehicle is, it will always need some fuel). This is unlike renewable energy sources, whose cost-effectiveness does not really depend on the initial energy use (e.g., installing solar panels on a wasteful home yields similar returns as installing the panels on an efficient home). Also, renewable energy can bring our use of fossil fuels down to zero, or close to it (AEA and REAP, 2016). The conceptual relationship between energy efficiency and renewable energy is shown in Figure 1.5. The "cost per unit of fossil fuels saved annually" represents the cost-effectiveness of the given measure. For example, the light bulb upgrade example in the Energy Efficiency section earlier shows savings of 110 kWh annually with the initial investment of US$3. So, it takes about US$0.03 of an initial investment to be saving 1 kWh of fossil fuel energy every year, which is very cost-effective and thus this measure would lie low on the graph.

Therefore, the most cost-effective approach toward eliminating the consumption of fossil fuels in the Arctic is to become very energy efficient first, up to the point where energy efficiency measures are starting to be too expensive compared to installing renewable energy sources (the intersect in Figure 1.5) and then produce the reduced amount of energy that is still needed from renewable sources. Energy efficiency is often called "the first step toward renewable energy."

Note that the above approach is a simplified explanation of complex issues that are beyond the scope of this chapter. The technical feasibility of eliminating fossil fuel use in the Arctic depends partly on the nature of renewable resources available. Unlike renewable energy sources that incorporate energy storage (such as conventional river hydro), eliminating the fossil fuel use in areas where only intermittent renewable energy sources (such as solar) are available is very challenging. Due to a large number of remote communities, integrating intermittent

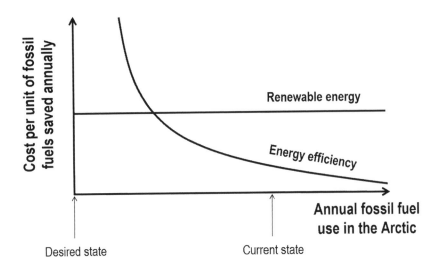

Figure 1.5 Comparison of energy efficiency and renewable energy from economic perspective.

Source: Authors.

renewable energy sources into isolated microgrids is an important research topic in the Arctic (ACEP, 2016).

Summary and conclusions

At present, fossil fuels still supply a majority of the energy used in the Arctic and generate a large amount of carbon emissions. This significant impact is partly due to Arctic-specific challenges, such as cold climate and long distances between communities, which are often small in size. Besides the vast use of fossil fuels, these challenges also pose other issues, such as energy insecurity and increased energy costs.

However, these challenges also present opportunities for implementing cost-effective energy efficiency measures and utilizing renewable energy technologies. Besides decreasing the Arctic's carbon footprint, some of the approaches developed in the Arctic serve as a model for other parts of the world. The Arctic has been seen as a pioneer in energy efficient construction techniques and also has a leading role in advancing renewably powered microgrid technologies.

The opportunity with the highest potential in the near-term future is energy efficiency. Energy efficiency, in general, provides the greatest reduction in fossil fuel use per monetary unit invested. However, the amount of energy reductions achievable through cost-effective energy efficiency measures has its limits. At the end, it is a highly energy efficient system powered by renewable energy sources that constitutes the vision for a low carbon Arctic energy system.

Acknowledgments

This work was done at the University of Alaska Fairbanks Bristol Bay Campus and the Cold Climate Housing Research Center. It was supported by the State of Alaska, the National Science Foundation and the National Institute of Food and Agriculture, U.S. Department of Agriculture. The authors would also like to thank the following individuals for making this chapter possible: Kristin Donaldson, Jennifer Centers, Gisele Arruda, Jack Hebert, Bruno Grunau, Michele Doyle, Dustin Madden, Robbin Garber-Slaght, Molly Rettig, Ron Ponchione, and others.

Notes

1 The authors were not able to find detailed historical or current energy information for the Russian Arctic.
2 The authors were not able to find detailed data for Greenland separately, and therefore, data for Denmark as a whole is shown.
3 Accurate and complete data for just the Canadian territories in the Arctic, that is, excluding the province of BC, was not available. Canadian data with the territories including the province of BC was complete. Therefore the authors decided that using the data that was bundled with BC was still a more accurate (less diluted) representation of Arctic Canada's energy consumption, rather than looking at Canada as a whole.
4 Some of this consumption is in the forms of derived heat and electricity. However renewable energy is a major input for the production of both. In this chapter, derived heat and electricity from renewable sources has been listed under the renewable energy category.
5 Different types of electricity costs exist, e.g., fuel cost versus cost to the customers. The cost to the customers is higher than the fuel cost because it also includes fixed costs (e.g., maintenance of power lines, insurance, etc.) distributed among the kWh sold. The fuel cost, as opposed to the cost to the customers, is used in this example to reflect the community impact (energy efficiency saves fuel, but not maintenance of power lines and other fixed costs).
6   Levelized Cost of Energy (LCOE, also called Levelized Energy Cost or LEC) is a cost of generating energy (usually electricity) for a particular system. It is an economic assessment of the cost of the energy-generating system including all the costs over its lifetime: initial investment, operations and maintenance, cost of fuel, cost of capital. A net present value calculation is performed and solved in such a way that for the value of the LCOE chosen, the project's net present value becomes zero.

(NREL, 2017b)

References

ACEP—Alaska Center for Energy and Power (2017) "Alaska Hydrokinetic Energy Research Center," http://acep.uaf.edu/programs/alaska-hydrokinetic-energy-research-center.aspx.

ACEP (2016) "Alaska Energy Technology Reports Overview Document," www.akenergy authority.org/Portals/0/Policy/AKaES/Documents/Reports/TechnologyDevelopment Needs.pdf.

ACEP (2013) "An Investigation of Solar Thermal Technology in Arctic Environments," http://energy-alaska.wdfiles.com/local-files/feasibility-of-solar-hot-water-systems/ KEA%20Solar%20Thermal%20Final%201-31-13.pdf.

AEA and REAP—Alaska Energy Authority and Renewable Energy Alaska Project (2016) "Renewable Energy Atlas of Alaska: A Guide to Alaska's Clean, Local, and Inexhaustible Energy Resources," http://alaskarenewableenergy.org//wp-content/ uploads/2016/07/RenewableEnergy-Atlas-of-Alaska-2016April.pdf.

Arctic Energy Alliance (2017) "Reducing the Use and Cost of Energy," http://aea.nt.ca/.

Askja Energy (2016) "The Energy Sector," https://askjaenergy.com/iceland-introduction/ iceland-energy-sector/.

Bernshtein, L. (1972) "Kislaya Guba Experimental Tidal Power Plant and Problem of the Use of Tidal Energy," in T. Gray and O. Gashus (eds.) *Tidal Power*. Springer, Boston, Massachusetts, USA, https://link.springer.com/chapter/10.1007/978-1-4613-4592-3_6.

Boyle, G. (2004) *Renewable Energy – Power for a Sustainable Future. Second Edition*, Oxford University Press, New York, USA.

CANSIM—Statistics Canada (2017) "Canadian Socioeconomic Database," www5. statcan.gc.ca/cansim/a01?lang=eng.

CCHRC—Cold Climate Housing Research Center (2017) "2016 Annual Report," https:// indd.adobe.com/view/c19a1100-023a-474c-9bf5-88b6378f49c8.

DOL—Alaska Department of Labor and Workforce Development, Research & Analysis Section (2017) "Annual Components of Population Change for Alaska, 1945 to 2017," http://live.laborstats.alaska.gov/pop/.

EIA/DOE—Energy Information Agency, US Department of Energy (2017) "State Energy Data System," www.eia.gov/state/seds/data.php.

Eurostat—Eurostat, the Statistical Office of the European Union (2018) "Energy Statistics – Supply, Transformation and Consumption," http://appsso.eurostat.ec.europa.eu/nui/ show.do?dataset=nrg_sankey&lang=en.

Freden, J. (2017) "The Swedish Recycling Revolution," https://sweden.se/nature/the-swedish-recycling-revolution/.

Gebre, S., Alfredsen, K., Lia, L., Stickler, M., and Tesaker, E. (2013) "Review of Ice Effects on Hydropower Systems," *Journal of Cold Regions Engineering*, vol. 27, no. 4, pp. 196–222.

Goldsmith, S., Pathan, S., and Wiltse, N. (2012) "Snapshot: The Home Energy Rebate Program," www.cchrc.org/sites/default/files/docs/HERP_snapshot.pdf.

Gunn, K. and Stock-Williams, C. (2012) "Quantifying the Global Wave Power Resource," *Renewable Energy*, vol. 44, pp. 296–304.

Helliwell, J., Layard, R., and Sachs, J. (2013) "World Happiness Report," http://unsdsn. org/resources/publications/world-happiness-report-2013/.

Hossain, Y., Loring, P., and Marsik, T. (2016) "Defining Energy Security in the Rural North—Historical and Contemporary Perspectives from Alaska," *Energy Research & Social Science*, vol. 16, pp. 89–97.

Kanellos, M. (2009) "Pelamis Wave Power Jettisons Its CEO, Rough Waters Ahead?," https://web.archive.org/web/20091003125522/www.greentechmedia.com/green-light/ post/pelamis-wave-power-jettisons-its-ceo-rough-waters-ahead.

Koponen, K., Sokka, L., Salminen, O., Sievanen, R., Pingoud, K., Ilvesniemi, H., Routa, J., Ikonen, T., Koljonen, T. Alakangas, E., Asikainen, A., and Sipila, K. (2015)

"Sustainability of forest energy in Northern Europe," www.vtt.fi/inf/pdf/technology/2015/T237.pdf.

Lee, M. and Reinhardt, G. (2003) *Eskimo architecture: dwelling and structure in the early historic period*, University of Alaska Press, Fairbanks, Alaska, USA.

Losnes, A. (2016) "Saami Reindeer Herders Fight Wind Farm Project," *Arctic Deeply*, 20 April, www.newsdeeply.com/arctic/articles/2016/04/20/saami-reindeer-herders-fight-wind-farm-project.

Marsik, T. (2009) "Basic study of renewable energy alternatives for electricity generation in Dillingham/Aleknagik region," www.agnewbeck.com/pdf/bristolbay/Dillingham_Comp_Plan/Report_NushagakAlternatives.pdf.

Marsik, T. (2018) "Renewable Energy Resources," in C.A. Woody (ed.) *Bristol Bay Alaska: Natural Resources of the Aquatic and Terrestrial Ecosystems*. J. Ross Publishing, Plantation, Florida, USA.

McDonnell, F. (2016) "Strong wind forecast for six counties," *The Irish Times*, December 5, www.irishtimes.com/special-reports/ni-trade-investment/strong-wind-forecast-for-six-counties-1.2893716.

MEAE—Ministry of Economic Affairs and Employment (2017) "Energy Efficiency," http://tem.fi/en/energy-efficiency.

Mokveld, K. (2011) "Energy efficiency in Russian Industry," www.iea.org/media/pams/russia/EnergyefficiencyinRussianIndustryv2022.pdf.

Mooney, C. (2015) "The remote Alaskan village that needs to be relocated due to climate change," *Washington Post*, February 24, www.washingtonpost.com/news/energy-environment/wp/2015/02/24/the-remote-alaskan-village-that-needs-to-be-relocated-due-to-climate-change.

Myles, P. (2017) *Maritime Clusters and the Ocean Economy: An Integrated Approach to Managing Coastal and Marine Space.* Routledge, London, Great Britain, https://books.google.com/books?id=OdkrDwAAQBAJ&pg.

NASA—National Aeronautics and Space Administration (2013) "Arctic Amplification," https://earthobservatory.nasa.gov/IOTD/view.php?id=81214.

NRCan—Natural Resources Canada (2017) "Comprehensive Energy Use Database," http://oee.nrcan.gc.ca/corporate/statistics/neud/dpa/menus/trends/comprehensive_tables/list.cfm.

NREL—National Renewable Energy Lab (2017a) "Arctic Renewable Energy Atlas," https://maps.nrel.gov/area-atlas.

NREL (2017b) "Simple Levelized Cost of Energy (LCOE) Calculator Documentation," www.nrel.gov/analysis/tech-lcoe-documentation.html.

Robarts, S. (2016) "Norway to build Europe's largest onshore wind power project," *New Atlas*, February 26, https://newatlas.com/fosen-vind-largest-wind-power-project-europe/42059/.

Roberts, B. (2017) "Solar Photovoltaic Advancements – The Arctic Energy Summit," http://arcticenergysummit.com/files/arcticenergysummit_mosey_2-20170928032619.pdf.

Ristinen, R., Kraushaar, J., and Brack, J. (2016) *Energy and the Environment. Third edition*, John Wiley & Sons, Hoboken, New Jersey, USA.

Sorensen, B. (2004) *Renewable Energy – Its Physics, Engineering, Environmental Impacts, Economics and Planning. Third Edition*, Elsevier Academic Press, Burlington, Massachusetts, USA.

Statistics Denmark (2018) "FT: POPULATION FIGURES FROM THE CENSUSES," www.statbank.dk/statbank5a/SelectVarVal/Define.asp?MainTable=FT&PLanguage=1&PXSId=0&wsid=cftree.

Statistics Finland (2017) "Population Statistics," www.stat.fi/tup/vl2010/index_en.html.

Statistics Iceland (2017) "Population – key figures 1703–2017," http://px.hagstofa.is/pxen/pxweb/en/Ibuar/Ibuar__mannfjoldi__1_yfirlit__yfirlit_mannfjolda/MAN00000.px.

Statistics Norway (2017) "07459: Population, by sex and one-year age groups. 1 January (M) 1986–2017," www.ssb.no/en/statbank/table/07459?rxid=029ca8e5-0b3b-4f4f-b36c-008e6d49023f.

Statistics Sweden (2017) "Population and Population Changes 1749–2016," www.scb.se/en/finding-statistics/statistics-by-subject-area/population/population-composition/population-statistics/pong/tables-and-graphs/yearly-statistics-the-whole-country/population-and-population-changes/.

Swedish Energy Agency (2017) "Programme for Energy Efficiency in Energy Intensive Industry (PFE)," www.motorsummit.ch/sites/default/files/2017-06/430_ms14_franck_0.pdf.

The Energy Efficiency Partnership (2013) "Alaska Energy Efficiency," www.akenergyefficiency.org/.

Tidal Energy Today (2017) "Seabased cancels Sotenäs wave park expansion. Continues R&D," https://tidalenergytoday.com/2017/10/30/seabased-cancels-sotenas-wave-park-expansion-continues-rd/.

Part II

Modalities of energy in the Arctic

2 Renewable energy for the Arctic

New perspectives

Gisele M. Arruda, Feb M. Arruda and
Julianna Mae Hogenson

Introduction

Chapter 2 provides a reflection on the Arctic renewable energy systems under the lenses of sustainable development. It addresses the new perspectives of renewable energy production and use, within the intercultural perspectives important to the development of a sustainable energy system in the Arctic. It is important to highlight that 'energy' is considered here to be a geographical concept, as its production and use presents different dynamics in distinct geographical contexts especially in the Arctic. The chapter concludes by arguing that the Arctic socio-environmental systems will benefit from the innovative way of dealing with the local energy system.

This research was originally conceived by the perception acquired through expeditions made to circumpolar areas in Canada, Greenland, Iceland, Alaska and Norway and it is a fruit of direct contact with experts, local communities and local energy production areas. The experience acquired from the fieldwork added value in the sense that it confirmed the perspectives presented in this chapter and provided access to the data collection. This data collection resulted in views, findings and conclusions regarding the current and future status of sustainable energy system in the Arctic. The text is structured in a way that the reader is shown the current energy production status in the Arctic specifically in Alaska, Canada, Greenland, Iceland and Norway with an important counterpoint about Russia. This scenario gives us a picture of an on-going process of energy transition in which the traditional oil and gas reserves are being explored and the remaining fossil unexplored reserves are considered of good potential by southern nations. In the same Arctic landscape, there are renewable energy projects being initiated, however there are a variety of methods that can be implemented to continue developing and dealing with the different energy production systems.

The chapter aims at expanding the understanding of the Arctic energy landscape and how energy production is linked to local social processes and ultimately to sustainable development. The Arctic is a complex geographical area where geophysical and climatic aspects are entangled with environmental and social factors, by showing that the balance among economic, social and environmental aspects are quintessential to the region's prosperity and survival. The chapter also brings clarity about energy production/use, impacts and adaptability to transitions and the relationship between these points in order to propose a reflection on the Arctic renewable energy leadership, energy technological solutions that can be applied in other areas of the globe, and the importance of clean energy for the region's sustainable future. The potential for energy development varies on a regional scale, but is beginning to be taken into consideration for the development of local communication and knowledge bases. The aim is to focus on Arctic areas where the transition is operating and generating a completely different local perspective.

Energy, climate and the Arctic

The end of the Cold War has triggered a new process for the Arctic region as it became less militarized and a new open space was created for entrepreneurship and economic development stimulated by political support on industrialization based on the abundance and the attractive prices of local commodities. These factors increased interest in industrial and maritime activities in the region representing an incipient and unregulated reality for the local context. Industrial and maritime activities are synonymous of economic development, but they also represent additional socio-environmental risks apart from the risks imposed by climate change.

The Arctic plays a key role in global climate. As the ice retreats the current challenges concerning the environment, maritime safety, tourism and oil and gas activity will intensify the effects of climate change on Arctic ecosystems and communities even more. Climate change and modernization have thus become two intrinsically linked forces that severely alter the context in which the local population of the region sustain a livelihood (van Voorst, 2009; Arruda and Krutkowski, 2017).

The development of industrial activities in the Arctic and sub-Arctic regions is not a new venture. Over centuries, alongside fishing, shipping, tourism and mining, oil and gas represented an important industrial activity due to the large fossil fuel resources concentrated in areas of Alaska North Slope, Beaufort Sea, Barents Sea, Amerasian Basin and East Greenland Rift Basin.

The Arctic is the home of an unexplored cluster of natural and human resources. It is known that 20 per cent of the world's energy resources in terms of fossil fuels like oil, gas and coal are in the Arctic (USGS, 2008). The US Geological Survey (USGS) estimates that 23 billion barrels of oil and 108 trillion cubic feet of natural gas offshore are recoverable from the Outer Continental Shelf (USGS, 2008).

More than 70 per cent of oil is estimated to lie in just five geologic provinces: Arctic Alaska, The Amerasia Basin, the East Greenland Rift Basins, the East Barents Basins and the West Greenland – East Canada sector. More than 70 per cent of natural gas is estimated to be in only three provinces: The West Siberian Basin, the East Barents Basins, and Arctic Alaska along the North Slope (Perry and Andersen, 2012, p. 14).

The Arctic region comprises vast amounts of extractive energy resources, most in offshore areas, which have been unreachable for a long time. Climate change and associated ice melting are making these areas more accessible for new extractive purposes. The main areas of extractive potential are the North Slope in Alaska (US); East Siberia, West Siberia, The Timan-Pechora area, South/North Barents (Russia), the area East of Norway and East Greenland. Other important areas include the Vilkitsky sector, the Laptev Sea, the Vilyuy sector, the Khatanga Sea and the North Kara Sea (Russia); the West Barents Sea, North of Russia; Denmark and Norway; the West Greenland area; the Sverdrup area north of Canada; and the Beaufort-Mackenzie area north of Canada and Alaska (AMAP, 2007, p. 10).

Oil and gas potential is significant in Arctic Alaska. On Alaska's North Slope, the nation's largest oil field, Prudhoe Bay, has been in production for several decades. Oil has been produced from the Beaufort Sea Ocean Continental Shelf (OCS) since the early 2000s, and the Arctic OCS potential for production of additional oil and gas resources is significant. Beyond petroleum potential, this region supports unique fish and wildlife resources, ecosystems and indigenous people who rely on these resources for subsistence. While the potential for, and interest in, energy resources is clear, there is significant public discourse over the ability to develop oil and gas resources safely, to understand environmental and social consequences of any development, and to implement effective impact prevention and mitigation strategies. Such discourse often revolves around different views on the sufficiency of the scientific information available to evaluate energy development options and to understand environmental sensitivity in this energy frontier area (Holland-Bartels and Pierce, 2011, p. 4).

Challenges exist in the Arctic fossil fuels exploitation ventures, and are generally linked to technical, environmental, economic, commercial and social aspects. A few examples of challenges in energy exploitation include: Remote locations with sensitive environments, challenges with icing on equipment, equipment safety design, high cost operations, lack of safety standards and extreme impacts on Arctic communities. However, the eight Arctic nations have an invaluable opportunity to make genuine and effective changes in how the Arctic is viewed and to create a new Arctic energy system that will reflect a new development model for the future in the whole Arctic region. Presently, despite the undeniable challenges to tackle for the high levels of energy poverty, social sensitivity, workforce unpreparedness and environmental fragility, we can envisage a great potential for sustainable energy prosperity (Arruda, 2015, p. 500).

Considering the operational and socio-environmental impacts of hydrocarbons production and use, but above all the climatic challenges as well as the Paris COP21 Emission Reductions Targets and their implications for Arctic Cryosphere, it is imperative to design and achieve a sustainable Arctic energy system. This is achieved through renewable energy production and use to secure a path to promote environmental resilience, energy self-sufficiency, energy independence and higher local living standards under a more decentralized system.

It is important to have in mind what sustainable energy means. According to *Collins English Dictionary* (2014) the term 'sustainable' means 'capable of being sustained', 'capable of being kept in existence' or 'capable of being maintained at a steady level without exhausting natural resources or causing severe ecological damage'. Considering the lack of capability to sustain the fossil fuel situation due to limited reserves, the exponential growth in energy demand and the severe socio-environmental impacts of oil extraction, renewable energy seems to be a significant component for the future of the Arctic region. It means producing and using energy in a sustainable way using practices that can be kept in existence, without compromising future generations. We get closer to energy sustainability by reducing the consumption of fossil fuels through energy efficiency and renewable energy. Some of the most ambitious renewable energy projects in the Arctic proved that energy efficiency needs to be hand-in-hand when assessing costs, risks and benefits.

Technology, policy innovation and energy governance will be crucial to achieve the continued decoupling of carbon emissions with a certain level of economic growth. Rethinking the current energy systems involve weakening the link between economic growth and carbon emissions as the base for future policies and technologies to be implemented. Modernization, according to the new paradigm, will require a shift in the composition of primary energy supply, coupled with pro-active energy efficiency policies that substantially reduce demand.

The Arctic energy landscape

There is no better time to review what is meant by 'Arctic' energy. Never before has there been so much technical and financial resources to engage in the dialogue on how energy systems can be developed in the Arctic region of the twenty-first century. In order to begin this reflection to expand the understanding on the Arctic energy landscape and, consequently, to be able to map the resource potential of the region, it is useful to consider two distinct energy sources: extractive and renewable.

Extractive resources comprise oil, gas, coal, natural gas from coal, methane hydrates, while renewable energy sources comprehend three basic types, solar energy (direct solar and wind), tidal energy and geothermal energy. A broad perspective of renewable sources involves solar, wind, geothermal, hydro and biomass, as seen in Chapter 1.

The Arctic energy landscape is responsible for nearly 20 per cent (USGS, 2008) of the world's energy extractive resources. However, apart from these more traditional energy sources, the Arctic presents a substantially vast unexplored potential of renewable energy resources that will help to deal with the projected local and global electricity demand within the 30 years ahead.

It is not an exaggeration to state that the High North region can be the energy power station of the world by providing clean sources of new energy for the world. The only aspect that needs clarification is to know what kind of energy source should be further developed. Extractive energy development is, in general, a commercial activity of interest to the Arctic nation's regional energy security and to the non-Arctic new energy demands for fossil fuels. Climate change effects, while opening new sea and land routes, are also changing the Arctic energy approach. However, what interests us most in this research, mainly considering the ambitious climate targets of COP21 (2015 Paris Climate Conference), is how to replace traditional fuel sources, to lower the local energy costs of electricity generation as well as to increase the quality of life of the local remote communities.

Renewable energy sources that have been considered in governmental plans both locally, regionally and internationally (IEA, 2013, p. 18) are also part of this energy landscape and could substantially support the global energy transition process. Energy plans materialized in local projects being presently developed in several scattered parts of the Arctic, demonstrate the start of an important energy transition trend, a proper shift in the mind-set with the potential to transform the Arctic energy matrix and, consequently, the adjacent nations energy systems.

Alaska has more than 175 remote village populations that rely almost exclusively on diesel fuel for electricity generation and heating (Goldsmith 2008, AEA 2014). Currently, Alaska generates 52 per cent of its electric power from natural gas produced in Cook Inlet near Anchorage. Hydropower is the next largest contributor, providing 24.9 per cent of electric generation, followed by oil at 13.6 per cent, coal at 6.1 per cent and wind at 2.5 per cent. While natural gas is prevalent in more urban areas, power in most of rural Alaska is generated by diesel fuel (59 per cent), followed by hydro at 29 per cent, natural gas at 8 per cent, and wind at 5 per cent (REAP, 2016). The consequence of this matrix model is that the cost of energy in most Alaska communities is as high as the cost of heating, which makes it difficult to think about sustainability when people are cold. The poorest households spend up to 47 per cent of their income on energy, representing more than five times their urban neighbours (The Alaska Arctic Council Host Committee, 2016, p. 14). Alaskans know the price of the world's most important commodity. Despite producing oil locally, no discount is provided for the diesel and gasoline that comes from refineries to rural communities via ice road, or air transport, which also contributes to the high local prices at the pump. Due to these and other factors, electricity generated by diesel fuel in some rural communities can be $1.00/kilowatt-hour (kWh) or more, which is more than eight times the national average of $0.12/kWh (AEA, 2014, EIA, 2014).

Local policy and legal framework materialized by the Senate Bill 138 issued by the Alaska State Legislature defined a new energy perspective towards a plan called the 'Alaska Affordable Energy Strategy' (AEA, 2016a) to effectively deliver affordable energy to areas without direct access to the Alaskan grid. The focus of this plan has been to evaluate the specific energy needs to determine the appropriate technology needed to be implemented at a local level in order to create adequate solutions to make energy more affordable. This is mainly for those remote, rural communities that lack a direct connection and are not supported by the Alaskan grid, the North Slope Natural Gas Pipeline.

'Alaska Affordable Energy Strategy' encouraged the integrated approach to sustainable energy in Alaska as an attempt to meet the social needs of its population mainly in rural Alaska. Since 2008, renewable energy and energy efficiency have been stimulated by a series of important regulations issued by the Alaska Energy Authority (AEA). One example is the House Bill 152 that established the Renewable Energy Fund (REF) providing *c.* $259 million in renewable energy projects across Alaska representing an energy saving of approximately 22 million gallons of diesel fuel a year (AEA, 2016b). In 2010, two other important bills were passed, Senate Bill 220 and House Bill 306. The first one established the Emerging Energy Technology Fund (EETF) with the aim at stimulating innovation and new technologies not tested in Alaska. House Bill 306 was fundamental to set goals oriented to convert 50 per cent of the state's electricity from renewable resources by the year 2025 as well as reducing energy use at the rate of 15 per cent per capita by 2020 by investing more than $360 million for home weatherization in new green-building technology, passive house standards and green construction programs (AEA, 2017).

After a process of initial consultations, the technologies adopted in Alaska included wind power, energy storage, diesel engines, hydroelectric systems, biomass, solar photovoltaic, heat pump and organic Rankine cycle (ORC), as well as electrical transmission and integration technologies that will be explored in more detail in Chapter 5.

The year 2012 visibly represented a game change in relation to wind generation in Alaska as important wind projects were implemented in locations like Kodiak with installed capacity of six turbines generating approximately 9 MW, Fire Island with a total of 11 turbines generating approximately 18 MW and the largest project Eva Creek with a total of 12 wind turbines with an installed capacity of approximately 25 MW at an investment of $3,780 per installed kW (REAP, 2016).

Another important energy resource that experienced a significant increase in installed capacity since 2012 was solar photovoltaic. The prospects of solar energy being part of the Alaskan energy system became a reality supported by local technology oriented to explore the solar potential despite the usual challenges like the reduction in electrical resistance associated to the low temperatures, albedo and light reflection that reduces the solar panels life cycle. The National Renewable Energy Laboratory (NREL) in conjunction with local energy centres have reported current solar photovoltaic installation ranging in size from 2.2 kW in Ambler to 50 kW in Galena (Whitney and Pike, 2017).

In Canada, the energy scenario is similar to the Alaskan Arctic, as 257 of Canada's 292 remote off-grid communities are still powered by diesel-fired power plants (Natural Resources Canada, 2011). Communities like Nunavut, for example, are completely reliant on diesel fuel, and in the Northwest Territories diesel represents the most expensive source of energy thus making energy transition a real concern for the Canadian decarbonization agenda. Northerners use

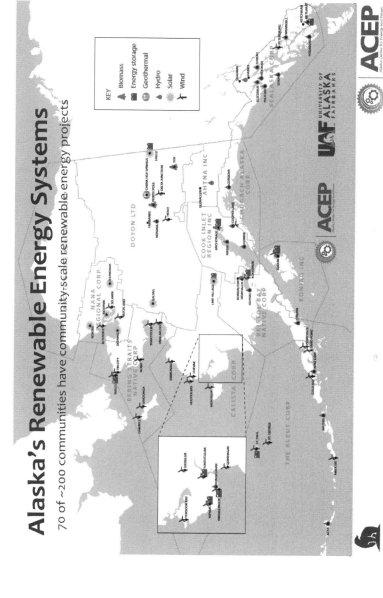

Figure 2.1 Renewable energy projects across Alaska.
Source: REAP (2016).

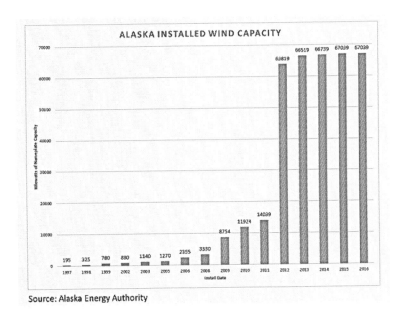

Figure 2.2 Alaska installed wind capacity.

Source: Alaska Energy Authority.

about twice as much energy as the national average, due to the long, cold and dark winters they endure. Being reliant upon diesel results in a large amount of greenhouse gases emitted in this process as the fuel is shipped over long distances, increasing the risks of local pollution.

Diesel reliance presents short and long-term implications for remote communities in the Canadian Arctic. Apart from the environmental impacts associated with oil spills, leakages, and inadequate storage causing soil and groundwater contamination, the high costs associated with electricity generation based on fossil fuel stimulate the creation of incentives to implement renewable energy projects. The Government of Canada's report entitled 'Status of Off-Grid Communities in Canada' emphasizes that the fuel prices are 'highly dependent on the mode of transportation to the delivery site' and the local communities (end-users) are not paying the true cost of their usage as figures can reach approximately $90 million annually considering the prices of fuel, added to the high transportation costs plus costs of supplying and storing diesel fuel, maintenance and operations (Natural Resources Canada, 2011; Advanced Energy Centre, 2015). As a result of fossil fuel logistics, the most isolated populations invariably suffer with serious environmental contamination risks.

Despite these risks, there has not been enough progress in developing renewable energy sources that could reduce these communities' reliance on diesel. Within the 80 communities in Canada's territories, 53 are, currently, still

Table 2.1 Diesel offsets for solar installations in Alaska

Village	Rated size (kW)	PV capacity factor (%)	2013 Community diesel efficiency (kWh/gal)	Average daily Solar performance Since installation (kWh)	Annual diesel offset (gallons)
Ambler	8.4	9	14.1	17.5	453
Ambler IRA	2.2	12	14.1	6.1	157
Kobuk	7.4	6	14.3	10.8	275
BSNC	9	9	16.2	37.3	840
Shungnak	7.5	7	14.3	12.4	316
Noorvik	12	6	12.4	17.6	518
Noatak	11.3	8	14.1	21.1	546
Deering	11.1	10	13.6	26.9	721
Selawik	9.7	11	13.9	25	656
Yuut Elitnuarviat (Bethel)	10	14	13.7	33.6	895
Kaltag	9.6	9	13	21.7	609
Galena	6.7	12	13.1	18.6	518
Ruby Washeteria	5.4	10	13.4	12.8	348
Ruby Health Clinic	5.5	8	13.4	10.8	294
Manley	6	9	12.5	12.3	359
Nenana	4.4	12	GVEA	12.5	–
CCHRC	8	15	GVEA	29.7	–

dependent on diesel fuel (Senate, Standing Committee on Energy, the Environment and Natural Resources, 2015). The high cost for imported hydrocarbons encourages renewable energy. With hydropower being the preferred option despite the large upfront investment and the significant costs of maintenance in case of photovoltaic and wind systems, technology costs proved to be a serious barrier to the adoption of renewable energy resources in the remote Canada.

The Canadian government is committed to implementing innovative energy solutions based on a holistic approach comprising of three viewpoints: policies, technology and local capacity. This approach is proving to be positive to engage stakeholders, like provincial governments, industries and communities, into an action-oriented planning to overcome the systemic challenges of developing renewable energy sources in the Canadian Arctic. According to the Advanced Energy Centre (2015), the implementation plan for renewable energy projects must follow these parameters because solely developing technical or policy solutions, without developing capacity within communities and utilities, will not sustainably advance innovation within local energy systems. Suppliers must consider key policy drivers and local social implications of technology, while building capacity within the local utility and communities for further technology adoption.

Since 2015, the Canadian government has demonstrated a clear trend to support a clean energy future for remote communities articulating collaboration from the public and private sectors to address the challenges preventing renewable energy systems to thrive in remote communities. It is the beginning of the transition from diesel generation to renewable energy and there is a significant demand for expertise, innovative energy products and integrated solutions. It involves a better understanding of socio-techno systems that will be further discussed in Chapter 5.

A more progressed renewable energy geographical area is the Nordic region, comprising of the states of Denmark, Finland, Iceland, Sweden, Norway, Greenland, the Aland Islands and Faroe Islands. Nordic countries, otherwise, have been at the forefront in achieving carbon-neutral targets in relation to energy systems in the last two decades. Several components of their energy systems have been successfully decarbonized, implying decoupled Greenhouse Gases (GHGs) emissions from Gross Domestic Product (GDP) growth. The carbon intensity of Nordic electricity supply was around 59 grammes of CO_2 per kilowatt-hour (gCO_2/kWh) in 2013, already at the level the world must reach in 2045 to realize the global 2°C Scenario (OECD/IEA *et al.*, 2016).

The integration of the grid has already been adopted in the Nordic electricity, thus acting as an important factor that has contributed to decarbonization. This was achieved by allowing hybrid systems to operate the balance between the 'peak and low' periods with the collaboration of neighbour countries' integrated renewable energy production.

The Nordic countries share a modern organized cooperation model and their close relationship is reflected in the internal and external energy grid integration. One example of the Nordic energy integration is Denmark and the fact that it represents the largest share of wind power converted into electricity generation in the world is a result of the balance provided by the hydropower system in Norway and Sweden. The Nordic renewable energy installed capacity results from an internally and externally connected grid interlinked through technological transmission lines and energy efficient electrical system integration programs.

The Nordic countries innovated their energy matrix due to the optimized use of the significant availability of renewable energy resources, mainly hydro and wind power. Within the Nordic communities, they share an innovated vision of the value of clean energy sources, by exchanging and sharing their energy surpluses. Currently, the power grids in Denmark, Norway, Sweden and Finland are integrated and interconnections also link the region to the Baltic States, Germany, Poland, the Netherlands and, in the near future the UK.

The Nordic cooperation envisages a stronger integration with other energy markets by altering the focus towards innovation and development of energy technology and climate solutions. An important strategic plan aligned to this viewpoint is the 'Nordic Solutions to Global Challenges' (Norden.org, 2017), a 2017 initiative with focus on promoting renewable energy, creating sustainable cities, integrating energy markets, and reforming fossil fuel subsidies schemes. It reassures the value of Nordic renewable energy resources.

Another remarkable example is the Icelandic energy sector that will be discussed in more detail in Chapter 3. It is unique in many ways, not only because it has abundant reserves of renewable energy in the form of hydro and geothermal resources, but due to the large share of renewable energy in the total primary energy matrix. In Iceland, 99.9 per cent of electrical production and 99 per cent of space heating is produced from renewable sources, thus achieving more than 73 per cent of the national mandatory target for 2020 from Article 3(1) of Directive 2009/28/EC (Ministry of Industries and Innovation, 2014).

The Icelandic government is still managing the transition as it has specific ambitions to decarbonize the transport and fisheries sector, which is dependent on imported fossil fuels. Policies are still required to increase the renewable energy share, to improve energy security, to reduce greenhouse gas emissions, to create environmental benefits, green jobs and global competitiveness. An energy efficiency plan is also in course to achieve carbon neutrality in households and commercial buildings. The plan contemplates a specific European Union regulatory framework setting standards for labelling and eco-design, oriented to provide clear information about energy consumption of appliances.

In Russia, the energy matrix is predominantly based on fossil fuels. Arctic Oil and gas drilling projects receive ample support from the government. However, since 2016 through the Arctic Council Project Support Instrument (PSI) and a partnership with the Nordic Environment Finance Corporation (NEFCO, 2017) a comprehensive environmental work has received financing for 76 environmental projects involving pollution prevention, hazardous waste management, climate change mitigation, energy efficiency and clean energy.

Projects aiming to mitigate black carbon and eliminate the use of mercury bulbs involved the installation of photovoltaic panels and energy storage equipment in remote off-grid communities of the Republic of Karelia like Justozero, Kimovaara, Lindozero, Voinitsa and Vozhmozero. Another important initiative happened in the village of Lovozero in the Murmansk Region, where light-emitting diode (LED) lights replaced mercury bulbs and a hybrid wind-diesel generator was installed to reduce emissions of black carbon and provide a better share of renewable energy. Other mappings are currently being produced with the intent to phase down diesel power plants in Dolgoshcheliye and in Northwest Russia.

The environmental mapping carried out by NEFCO and PSI is also connected to the Nordic countries' bilateral environmental cooperation. The Norwegian environmental programme in Archangelsk Oblast concerns the areas of oil pollution and the correct management of solid waste. These joint initiatives have been developed along 2016 and 2017 in order to minimize and prevent pollution in the Barents Region.

Innovative energy solutions for the Arctic – glacial meltwater power in Greenland

The Arctic energy landscape presents several examples of innovative energy solutions like the geothermal power in Iceland and microgrid usage in Arctic North America, respectively approached in specific chapters of this book. This section outlines the extraordinary energy solution in course in Greenland.

The energy system in Greenland has been historically characterized by the lack of integration, for geographical reasons, and by the imported fossil fuels as a primary source of energy. It is a region accounting for approximately 55,000 people distributed in six relatively big towns, 11 small towns and approximately 60 scattered settlements with the majority of the population living on the West Greenland coast.

In the last decade, considerable efforts have been made to restructure the energy system in Greenland towards a more sustainable matrix consisting of approximately 60 per cent of Greenland's energy being produced by renewable resources (Government of Greenland, 2015). Among recent advancements there are five hydroelectric power plants and six waste incineration plants operated by municipalities providing heating in several Greenlandic towns. Since 1 January 2014, renewable energy generators have received a subsidy based on the cost of having to supply the equivalent amount of energy by non-renewable means.

The main challenge in Greenland is the lack of an efficient energy transmission system, which makes it difficult to use the energy surplus or rely upon intermittent energy sources such as solar and wind. This creates the need to have a backup energy system in every community commonly based on diesel fuel.

Greenland reveals relevant examples of innovative approaches to energy production that contribute to increasing the reliability and sustainability of Greenland's energy system as well as local development. Recent initiatives include turning fish residues and other wastes into district heating in Sisimiut, autonomous use of renewable energy in Saarloq and generating hydropower using glacial meltwater in Ilulissat. Despite the importance of the two first projects, our focus resides on the Ilulissat glacial meltwater.

Ilulissat is a touristic site with different characteristics from other Greenlandic locations. It is the home of the Ilulissat Icefjord or, in Greenlandic, Ilulissat

Kangerlua with its eastern end in Jakobshavn Isbrae glacier or Sermeq Kujalleq as the most productive glacier generating 35 cubic kilometres of ice a year at an extreme high velocity of 19 metres a day (Bennike *et al.*, 2004).

The glacial meltwater is a mega hydropower project. It consists in a power plant in the permafrost fed by the discharge from natural glacial lakes where meltwater is channelled through 200 m down the permafrost via a transfer tunnel controlled by valves to underground turbines located at a 400 m powerhouse inside the mountain. The powerhouse is at sea level and the meltwater from the turbines empties into a tail race tunnel which again opens into the sea.

Ilulissat Icefjord is a World Heritage Site by the United Nations Educational, Scientific and Cultural Organization (UNESCO) since 2007, but an area of international research and scientific applications serving national and local interests. In terms of energy innovative application, this area supports a new hydropower plant installed beneath the permafrost and located in the Arctic circle at 50 km from Ilulissat reflecting one of the most ambitious and fully automated projects ever designed considering that it addresses, in a high technologically way, the severe risks imposed by low temperatures of minus 40°C to operate it. This is the state-of-the-art Arctic energy project showing how diverse and promising is the renewable energy ambition for the future of the Arctic.

Conclusions

The operational and socio-environmental impacts of hydrocarbons production and use, but above all, the climatic challenges and their implications for Arctic

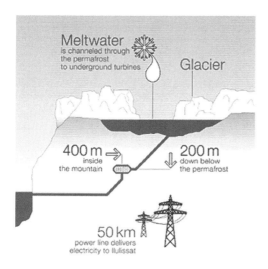

Figure 2.3 Ilulissat Hydroelectric Project. Adapted from ABB infographic Greenland hydropower.

Source: ABB Communications (2012) (used with authorization).

Cryosphere, have been imperative drivers to redesign a sustainable Arctic energy system.

The new perspective is achieved through renewable energy production and use to secure a path to promote environmental resilience, energy self-sufficiency, energy independence and higher local living standards under a more decentralized, flexible and efficient system. The Arctic renewable energy revolution is intrinsically driven by social needs that are transforming the Arctic energy landscape according to a continuous stream of socio-technical solutions with the intent to reduce the production and use of fossil fuels.

A variety of environmental challenges will require a new generation of solutions and investments in facilities to expand, adapt and integrate grids according to efficient modes of production, and new level of stakeholders' interaction.

Finally, it is extremely important to create a secure and sustainable Arctic energy system applying long-term strategies accompanied by effective investments in the supply, conversion, transmission, and storage side. This is certainly a case study of technology and innovation to reflect and apply worldwide.

References

ABB Communications (2012) 'Clean Sustainable Energy for Greenland', www.abb.com/cawp/seitp202/b08e¬a3b92dc74ac8c1257aaf0047543c.aspx, accessed 1 October 2017.

ACEP and AEA (2016) 'Alaska Energy Technology Reports-Overview Document. Documentation of Alaska – Specific Technology Development Needs in Support of the Alaska Affordable Energy Strategy', www.akenergyauthority.org/Portals/0/Policy/AKaES/Documents/Reports/TechnologyDevelopmentNeeds.pdf?ver=2016-08-08-152005-117, accessed 20 October 2017.

Advanced Energy Centre (2015) 'Enabling a Clean Energy Future for Canada's Remote Communities', December 2015, Advanced Energy Centre, Discussion Brief, Ontario, Canada.

Alaska Energy Authority (AEA) (2014) 'Power Cost Equalization Program: Statistical Data by Community', Reporting Period: 1 July 2012 to 30 June 2013. Issued February 2014, www.akenergyauthority.org/Content/Programs/PCE/Documents/FY13Statistical RptComt.pdf, accessed 11 November 2017.

Alaska Energy Authority (AEA) (2016a) 'Alaska Affordable Energy Strategy: A Framework for Consumer Energy Sustainability Outside of the Railbelt', www.akenergyauthority.org/Portals/0/DNNGallery/uploads/2017/1/23/AkAESES12317printfinalv2.pdf, accessed 20 October 2017.

Alaska Energy Authority (AEA) (2016b) 'Renewable Energy Fund', www.akenergyauthority.org/Programs/RenewableEnergyFund, accessed 20 September 2017.

Alaska Energy Authority (AEA) (2017) 'Renewable Energy Fund. Status Report and Round X Recommendations', www.akenergyauthority.org/Portals/0/DNNGalleryPro/uploads/2017/1/27/REF%20Round%20X%20Status%20Report.pdf, accessed 1 November 2017.

Arctic Council (2018) www.arctic-council.org/index.php/en/, accessed 20 December 2017.

Arctic Monitoring and Assessment Programme (AMAP) (2007) Arctic Oil and Gas 2007. Oslo, Norway, p. 10.

Arruda, G. M. (2015) 'Arctic Governance Regime: The Last Frontier for Hydrocarbons Exploitation', *International Journal of Law and Management*, Vol. 57, Iss. 5, pp. 498–521, p. 500.

Arruda, G. M. and Krutkowski, S. (2016) 'Arctic Governance, Indigenous Knowledge, Science and Technology in Times of Climate Change: Self-realization, Recognition, Representativeness', *Journal of Enterprising Communities: People and Places in the Global Economy*, Vol. 11, Iss. 4, pp. 514–528.

Arruda, G. M. and Krutkowski, S. (2017) 'Social Impacts of Climate Change and Resource Development in the Arctic: Implications for Arctic governance', *Journal of Enterprising Communities: People and Places in the Global Economy*, Vol. 11, Iss. 2, pp. 277–288.

Bennike, O., Mikkelsen, N., Pedersen, H. K., McCollum, G. and Weidick, A. (2004) 'Ilulissat Icefjord'. *Geological Survey of Denmark and Greenland*.

Collins English Dictionary (2014) 12th edition, accessed 26 October 2016.

Energy Information Administration (EIA) (2014) 'Electric Power Monthly: Table 5.6.A. Average Price of Electricity to Ultimate Customers by End-Use Sector, by State', www.eia.gov/electricity/monthly/epm_table_grapher.cfm?t=epmt_5_6_a, accessed 20 July 2017.

Goldsmith, S. (2008) 'Understanding Alaska's Remote Rural Economy', UA Research Summary No. 10, January 2008, Institute for Social and Economic Research, University of Alaska Anchorage, www.iser.uaa.alaska.edu/Publications/researchsumm/UA_RS10.pdf, accessed 20 December 2017.

Government of Greenland (2015) 'Hydropower and Renewable Energy', http://naalakkersuisut.gl/en/Naalakkersuisut/Departments/Natur-Miljoe-og-Justitsomraadet/Natur_-Energi-og-Klimaaf¬delingen/Energi/Vandkraft-og-vedvarende-energi, accessed 20 October 2017.

Holland-Bartels, L. and Pierce, B. (2011) 'An Evaluation of the Science Needs to Inform Decisions on Outer Continental Shelf Energy Development in the Chukchi and Beaufort Seas', Alaska, *U.S. Geological Survey Circular* 1370, https://pubs.usgs.gov/circ/1370/, accessed 20 July 2017.

International Energy Administration (IEA) (2013) 'Nordic Energy Technology Perspectives', www.nordicenergy.org/wp-content/uploads/2012/03/Nordic-Energy-Technology-Perspectives.pdf, accessed 1 November 2017.

Ministry of Industry and Innovation (2014) 'The Icelandic National Renewable Energy Action Plan for the Promotion of the Use of Energy from Renewable Sources in Accordance with Directive 2009/28/EC and the Commission Decision of 30 June 2009 on a Template for the National Renewable Energy Action Plans. Ref. Ares (2014)806315–19/03/2014, https://ec.europa.eu/energy/sites/ener/files/documents/dir_2009_0028_action_plan_iceland_nreap.pdf, accessed 1 December 2017.

Natural Resources Canada (2011) 'Status of Remote/Off-Grid Communities in Canada', www.nrcan.gc.ca/energy/oublications/sciences-technology/renewable/smart-grids/11916, accessed 20 September 2017.

Nordic Environment Finance Corporation (NEFCO) (2017) 'Arctic Council Project Support Instrument', www.nefco.org/sites/nefco.org/files/pdf-files/psi_manual_27.2.2017.pdf, accessed 20 December 2017.

Norden.org (2017) 'Nordic Solutions to Global Challenges', Nordic Cooperation, www.norden.org/en/theme/nordic-solutions-to-global-challenges/nordic-solutions-to-global-challenges, accessed 10 December 2017.

OECD/IEA, Nordic Energy Research, Technical University of Denmark, Ea Energianalyse A/S, VTT Technical Research Centre of Finland, University of Iceland, Institute

For Energy Technology, Profu Ab and IVL Swedish Environmental Research Institute 2016, Nordic Energy Technology Perspectives, IEA Publishing, p. 15.

Perry, M. C. and Andersen, B. (2012) 'New Strategic Dynamics in the Arctic Region: Implications for National Security and International Collaboration', The Institute for Foreign Policy Analysis, 25 May 2012, p. 14.

REAP (2016) 'Renewable Energy Atlas of Alaska. A guide to Alaska's Clean, Local and Inexhaustible Energy Resources', *Alaska Energy Authority*, April 2016.

Senate, Standing Committee on Energy, Environment and Natural Resources (2015) 'Powering Canada's Territories', YC26-0/412-14E-PDF, Ottawa, p. 9.

The Alaska Arctic Council Host Committee (2016) 'Environmentally Responsible Resource Use and Development in the U.S. Arctic', *The Alaska Arctic Council Host Committee*, October 2016, p. 14.

The US Geological Survey (USGS) (2008) 'Circum Arctic Resource Appraisal: Estimates of Undiscovered Oil and Gas North of the Arctic Circle', *The US Geological Survey*, USGS Fact Sheet 2008–3049.

The US Geological Survey (USGS) (2015) *Minerals Yearbook*, Volume I – Metals and Minerals, USGS.

van Voorst, S. R. (2009) 'I Work All the Time – He Just Waits for the Animals to Come Back: Social Impacts of Climate Changes: A Greenlandic Case Study', *Journal of Disaster Risk Studies*, Vol. 2, Iss. 3, pp. 235–252.

Whitney, E. and Pike, C. (2017) 'An Alaska Case Study: Solar Photovoltaic Technology in Remote Microgrids', *Journal of Renewable and Sustainable Energy*, Vol. 9, Iss. 6.

3   Geothermal energy development in the Arctic

Julianna Mae Hogenson

Geothermal energy

The civilized world continues to expand and grow into the farthest corners of the earth, furthering the need for sustainable energy resources. As the Arctic region becomes more accessible, a need arises for proper management of the resources located above and below the ground. For an energy source, geothermal energy presents itself as a viable option. Currently, geothermal energy is globally used for a variety of purposes including, but not limited to, electrical generation and heating. Geothermal should not be considered a "quick fix" solution to energy needs. Rather it is a long-term process resulting in an environmentally friendly, sustainable, and reliable energy source. A typical large-scale geothermal project takes 5–10 years to complete.

How geothermal works

Geothermal energy by definition is energy that is stored within the mantle and crust of the earth. The thermal energy available in the earth's crust lies within the natural fluids, in a variety of states, contained within fractures and pores situated in the upper boundary of the crust. When magmatic activity occurs within the crust, the thermal energy from the magma, in the form of heat, is transferred into the fluid thus creating hydrothermal fluids. The now hot fluids are extracted to harness the geothermal energy.

These hydrothermal fluids range from being in a liquid or supercritical fluid phase to a saturated or superheated steam vapor phase. The majority of the world's geothermal resources are liquid-dominated, with few vapor-dominated systems (Allis, 2000). The hydrothermal fluids being removed from the crust are part of what is known as a geothermal reservoir. A geothermal reservoir is a section within the crust characterized by an area of permeability containing a thermal aquifer and typically capped by a layer of rock with lower permeability (Blodgett, 2014).

According to calculations done in 1998 by Wright, there is approximately 1.3×10^{27} J of thermal energy within 10 km of the earth's surface. This is equivalent to consuming 3.0×10^{17} barrels of oil (Wright, 1998). The International Energy Agency (IEA) reports that in 2016, the global daily consumption of barrels of oil per day was 96 million barrels of oil per day, equaling 35×10^9 barrels of oil a year (IEA, 2017). This means that geothermal energy has the potential to provide energy for 8.57×10^6 years. Based on these numbers, it seems like geothermal could be a solution to solve a global energy crisis and provide heating and hot water. This ideal situation is very unrealistic.

There are a multitude of factors that affect the feasibility of harnessing geothermal energy. One key factor is the location of the geothermal reservoir and its spatial relation to consumers. It is dependent not only upon the geologic and tectonic classifications, but also on accessibility and reservoir productivity (Lund et al., 2008).

Types of geothermal resources

The three main characteristics that define a geothermal resource are temperature, depth, and permeability/porosity, but every geothermal resource will be unique. Ideal geothermal resources will be those that have high temperatures at shallow depth with a high degree of permeability and porosity. The three main classification categories for geothermal energy resources are: (1) high temperature resources where the reservoir temperature at a depth of one kilometer is greater than 200°C, (2) medium temperature resources where the reservoir temperature at a depth of one kilometer is between 100°C–200°C, and (3) low temperature resources where the reservoir temperature at a depth of 1 km is below 100°C. The possible uses of these types of resources based upon temperature are found in Figure 3.1.

Typically, high temperature resources are more suitable for electrical production. These resources are usually found where there are active volcanic systems. Such as where there are magmatic intrusions into the earth's crust, recent volcanic activity, or tectonic plate boundaries.

Medium temperature resources also can be used for power generation by using a binary power plant (Kwambai, 2014). Only the medium temperature resources that are greater than 150°C are suitable for power production and the lower temperatures are better utilized for direct use purposes (Kanoglu and Bolatturk, 2008).

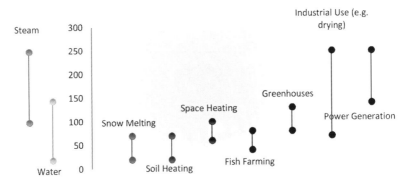

Figure 3.1 Geothermal resource utilization based upon temperature, in °C

Source: Data derived from Lindal, 1973.

Low temperature resources are utilized for direct use of a geothermal resource. They can be used for geothermal heat pumps, aquaculture heating, space heating, snow melting, industrial drying processes, greenhouses, and other uses.

These three types of resources can be found in three types of geothermal systems, (1) Natural Hydrothermal Systems, (2) Geopressured Systems, and (3) Enhanced Geothermal Systems (hot dry rock) (Tester et al., 2012).

Natural hydrothermal systems

Natural hydrothermal systems are usually located in or near areas of tectonic or volcanic activity. These are areas where the geothermal resource spontaneously produces fluid. The system is fed by rain, snow, or other forms of moisture that leach into the porous reservoir associated with a heat source. As the heated fluid comes to the surface, it becomes geothermal surface features, such as a hot spring, fumarole, solfatara, and/or geyser (Tester et al., 2012). Figure 3.2 will show a basic model of a natural hydrothermal system.

The two main types of hydrothermal systems are vapor-dominated and liquid-dominated. Vapor-dominated systems occur when steam and non-condensable gases separate from the liquid phase and accumulate at the top of the reservoir. This is because the pressure in the geothermal reservoir allows for boiling to occur. The steam being produced is ideal for electrical production, as it typically is very high in temperature and can be used in low-pressure steam turbines. The steam does not contain as many dissolved minerals as the liquid phase, thus there is less corrosion, scaling, and damage to the turbines and piping systems (Tester et al., 2012). Very few high-grade vapor-dominated geothermal systems exist in the world, but the most significant

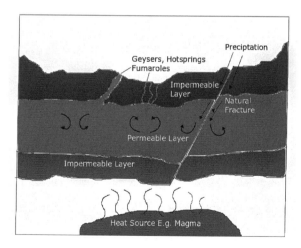

Figure 3.2 Natural hydrothermal system.

Source: Data derived from Tester et al., 2012.

are the Geysers system in California, the Matsukawa field in Japan, and the Larderello Field in Italy.

The liquid-dominated systems occur when the pressure in the system is high enough that it prevents boiling and thus the creation of the vapor cap in the upper portion of the system does not occur. Rather, the geothermal system contains hot water or a mixture of hot water and steam. The dissolved solids in the liquid create corrosion and scaling issues in the pipes and turbines. Mitigation strategies must be considered to counteract this. After the energy is extracted, the remaining fluid can still contain significant levels of dissolved solids and/or thermal heating potential. This fluid is usually dealt with through reinjection, evaporation ponds or direct use (Tester et al., 2012). Liquid-dominated systems are very common throughout the world and are used for electrical production and direct use in residential and industrial sectors.

Geopressured systems

Geopressured geothermal systems form when the brine in the reservoir is confined by an impermeable layer of caprock, typically located below interbedded sedimentary rock layers, such as shales and sandstones (National Renewable Energy Laboratory (NREL), 2010). These sedimentary layers create confining pressure on the impermeable layer of caprock over the geothermal system (Figure 3.3). The pore fluids trapped within the sedimentary layers become pressurized, up to 600 MPa (600 bar) and are very mobile with temperatures between 90–200+°C (Tester et al., 2012). Due to the nature of the host rock, for example shale, the fluid tends to contain dissolved natural gas and become saturated with methane. This is a problem for utilization, but presents an interesting opportunity to combine co-produced fluids from oil and gas wells into a geothermal project. The first geopressured geothermal power project in the United States, is still under development in Louisiana (National Renewable Energy Laboratory (NREL), 2010).

Enhanced geothermal systems

Two of the key necessities for a geothermal reservoir are porosity within the system and a fluid in which to extract heat from. Even if the rock at depth in a reservoir provides sufficient heat, if there is no permeability or fluid, a geothermal power system will not function properly. Areas where there is heat, but a lack of fluid and permeability are referred to as hot dry rock (HDR) or enhanced geothermal systems (EGS) (Tester et al., 2012).

Theoretically, an EGS power system could be feasible anywhere within the earth's crust, from 1 km to 10 km in depth where there is high enough temperatures. Typically, a EGS system is created by drilling to the required depth for sufficient temperatures, then hydraulically fracturing the rock. Once this is done, a second well is drilled within the fractured area. Between the two wells, water should be circulating from one well to the other in the fractured area (Figure 3.4).

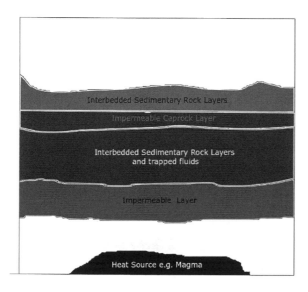

Figure 3.3 Geopressured geothermal system.

Source: Data derived from Tester et al., 2012.

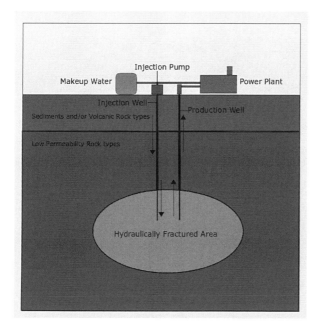

Figure 3.4 Enhanced Geothermal System (EGS).

Source: Data derived from Tester et al., 2012.

Although EGS systems are becoming more feasible, there are still unknowns associated with this type of geothermal system. Information regarding the fracturing process and its effects is lacking. Induced seismicity, thermal stress cracking, and chemical leaching are a few known side effects

Geothermal power systems

Power generation

The three major types of geothermal power plants are: (1) Dry-steam plants, (2) Flash-steam plants, and (3) Binary-cycle plants. Combined flash/binary plants are newer designs and not widely used in the geothermal industry.

Dry-steam plants are typically used in a vapor-dominated geothermal system with temperatures from 230–320°C (Eliasson et al., 2011). They are simpler in engineering design to other types of power plants. As of 2011, they consisted of 27 percent of the geothermal power plants in the world (DiPippo, 2012). This is despite the fact that dry-stream reservoirs are not as common throughout the world. In this design, the fluid is taken directly from the production well to the turbine. Figure 3.5 shows the turbine blades of a turbine at the Svartsengi power plant. More complicated systems can implement the use of a steam header, moisture removers, condensers, cooling towers and reinjection equipment.

Flash-steam plants use geothermal reservoirs at temperatures between 200–320°C (Eliasson et al., 2011). The two main types are single-flash and

Figure 3.5 Blades from a turbine at Svartsengi power plant, Iceland.

Source: Photo by author.

dual-flash power plants. Single-flash systems consist of 43 percent of installed geothermal power capacity globally and dual-flash power plants constitute 10 percent. The dual-flash system can generate 15–25 percent more power than a single-flash system (DiPippo, 2012). Due to the increased cost and complexity, the dual-flash system is not as popular. Single-flash systems operate when the fluid is sent through a flashing process and a separator before it reaches the turbine. The remaining working fluid goes through a condenser and a cooling tower (DiPippo, 2012). In Iceland, Krafla is a single-flash power plant producing 60 MW of power from two 30 MW steam turbines (Figure 3.6)

The dual-flash system reflects the same general outline as a single-flash system. The main difference stems from when the brine recovered from the separator is flashed again, but at a lower pressure than the primary fluid. Other differences will stem from the project design. For example, the placement of the separators and flashers at the wellhead location, then using two phase pipelines to transport the fluid to the powerhouse (DiPippo, 2012). The different designs will highly impact the cost and economic viability of a project. The Germencik geothermal power station near Izmir, Turkey, producing 45 MWe is an operating dual-flash system.

Binary-cycle plants, as of 2011, generate 6.6 percent of global geothermal power. A binary-cycle power plant will use resources with temperatures of 120–190°C (Eliasson et al., 2011). The efficiency of binary systems is significantly smaller in comparison to dry-steam and flash-steam power plants. The

Figure 3.6 Krafla, a 60 MW power plant located in Northern Iceland.
Source: Photo by author.

lower the temperature of the fluid, the harder it becomes to efficiently generate electricity (DiPippo, 2012). A basic binary-cycle will see the production wells regulated by pumps to ensure the pressure remains above its flashing point and there is a consistent flow rate. The geothermal fluid is brought to the boiling point in an evaporator, becoming a saturated vapor. Maintaining correct temperatures and pressures is key to reducing scaling and avoid the breakout of steam and non-condensable gases. The saturated vapor is sent to a turbine and then a condenser (DiPippo, 2012). Different types of binary cycles can have different efficiencies. A dual-pressure cycle with a dual-admission turbine will have a two-stage heating/boiling process so the average temperature difference will be lower. Other types include dual fluid cycles and Kalina binary cycles (DiPippo, 2012).

Direct use

For areas where there is not a high temperature geothermal system, direct use systems can be applied. When analyzing the possibilities for geothermal energy across a global scale, the electrical generation potential is significantly lower than the potential for direct use application. As of 2015, the total geothermal direct use installed capacity was 70,885 MWt. This is an increase of over 20,000 MWt, from 2010 (Figure 3.7) (Lund and Boyd, 2016). With advances in technology, the direct utilization of geothermal resources has expanded exponentially. As more ways are created to use the thermal potential of geothermal systems, it continues to grow across the globe.

As can be seen in Figure 3.7, the top three most common uses for direct use is geothermal heat pumps, space heating and bathing and swimming. In ten years, direct use for geothermal heat pumps has tripled in capacity, while direct use for space heating and bathing and swimming has almost doubled in capacity. Based upon this data, it can be assumed that as the industry continues to grow, the technology has become more established, thus more affordable.

Benefits and challenges of geothermal

As a source for an environmentally friendly and renewable energy, geothermal is more than qualified. There are several benefits to using geothermal energy over other renewable energy types. Most renewable energy sources, such as wind, solar and tidal are intermittent and cannot be a base load power source. Geothermal provides stable, consistent base load power. Another benefit is that typical geothermal system designs are proven technology and are commercially profitable projects (Gehringer and Loksha, 2012).

The land use per unit of power needed for a geothermal project is much smaller than other renewable resources. Making its environmental impact smaller and more mitigatable. The ability for the scaling up of geothermal projects with minor increases in land use reduces further environmental damage. For example, in Iceland, directional drilling is being implemented to drill multiple production wells from a single well pad (Figure 3.8). This in turn

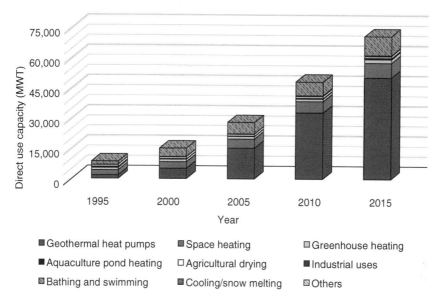

Figure 3.7 Geothermal direct use global installed capacity.

Source: Data derived from Lund and Boyd, 2016.

Figure 3.8 Directional drilling at Hellisheiði power plant, Iceland. There are now three wells drilled from the same well pad.

Source: Photo by author.

drastically reduces the environmental impact of scaling up the geothermal power plant capacity.

In comparison to traditional fossil fuels, such as oil, gas, or coal, the CO_2 and other greenhouse gas emissions from geothermal are almost negligible per unit of energy produced (Gehringer and Loksha, 2012). There will be geothermal resources that contain higher amounts of H_2S and CO_2 than others. Systems with high amounts of noncondensable gases (NCGs) are unsuitable for electrical production, as the NCGs will damage the turbines.

One of the disadvantages of geothermal is that the ideal geothermal reservoir is not readily found around the globe. Rather they are concentrated around volcanic systems. If the geothermal source is inaccessible or far away from the targeted consumer, the economic viability of the project decreases. It also increases the environmental impact of the project as more infrastructure is required for piping and transmission lines to access the consumer.

In non-volcanic areas, geothermal may still be viable, but it is more challenging to develop a functioning and economically viable project. The more unproven geothermal system types, such as EGS or geopressured systems, could be implemented in these areas. Unfortunately, they have higher risks and uncertainties. The technical side of the geothermal project tends to present challenges and difficulties for development. The phases of development for a geothermal project will be explained further on.

Geothermal energy is renewable if the reservoir is managed properly. As the fluid is extracted through production wells, the rate of extraction is often higher than the natural heat flow. This will result in cooling regions within the reservoir. Eventually, the stored thermal energy will regenerate through conduction from the source rock and convection from the geothermal fluids. The recovery time is typically greater than the rate of extraction. If this occurs, the reservoir will become depleted and the lower temperatures reduce the heat and power available from the reservoir. The cooling of production boreholes is a serious risk to the lifetime of a geothermal reservoir. Therefore, reservoir management strategies unique to each system need to be devised.

A method for recharge in the geothermal system is implementing the use of reinjection wells. These are wells located upstream in the geothermal reservoir, where the cooler, energy depleted fluid is pumped back down into the geothermal reservoir. It will recharge as it flows through system. Originally, reinjection wells were used to dispose of the waste water from the geothermal process, now they are a vital part of resource management.

Tracer tests are used to better understand the fluid dynamics of a geothermal power system and identify reinjection wells. This is done once a geothermal reservoir system has multiple production or exploration wells drilled and functioning. Tracer tests consist of injecting a chemical tracer into a geothermal well and then samples are taken over time at other wells in the geothermal field. These samples will determine the rate and amount of dispersal throughout the reservoir. The type of geothermal system will determine what type of chemicals can be used. In vapor-dominated systems, fluorinated hydrocarbons and sulphur hexafluoride are commonly used. For liquid-dominated systems, halides, radio-active tracers, fluorescent dyes, aromatic acids and naphthalene sulfonates are used (Axelsson, 2013). To ensure accuracy in tracer testing, the tracer chemical must not exist naturally within the reservoir.

The most significant benefit of completing tracer tests is the gathering of information on the reservoir structure and flow patterns. This information helps understand the flow of the reservoir between reinjection and production bore-holes (Axelsson, 2013). By understanding the recharge rate of a geothermal system, calculations can determine the capacity of the geothermal reservoir, create cooling predictions and help develop management strategies to prolong the lifetime of the reservoir.

Geothermal developments

According to the World Energy Council, the global installed capacity of geother-mal power was 13.2 GW in 2016. That is an increase by 315 MW from the 2015 power capacity (World Energy Council, 2016). By the end of 2016 the top five countries for geothermal capacity and generation for power are in order: 1. United States, 2. Philippines, 3. Indonesia, 4. New Zealand, 5. Mexico. The top five countries for geothermal capacity or generation for heating purposes are in order: 1. China, 2. Turkey, 3. Japan, 4. Iceland, 5. India. (REN21, 2017).

Geothermal power plant development phases

The typical development of a geothermal power plant project consists of eight phases (IGA Service GmbH, 2014; Gehringer and Loksha, 2012):

1   Preliminary survey
2   Exploration
3   Test drilling
4   Project review and planning
5   Field development
6   Power plant construction
7   Commissioning
8   Operation

Due to the uniqueness of each geothermal project, these phases will be developed differently on a project-to-project base. The general outline is applicable to most projects. As there is no set standard of global energy regulations, each country has its own set of legal restrictions that will determine the specifics of a geother-mal project. Earlier geothermal projects, such as the Geysers project in Italy and the Wairakei area in New Zealand, did not follow any systematic process. Now there are regulatory methodologies and techniques in place for most countries involved in the geothermal industry, e.g., Africa and Iceland. Throughout the first three phases of development, the risk for the project is the highest. As the

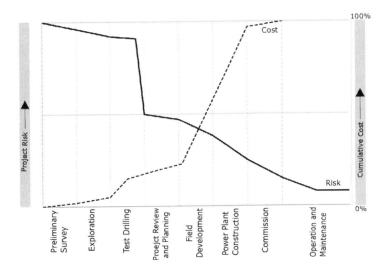

Figure 3.9 Geothermal project risk and investment cost.

Source: Data derived from Gehringer and Loksha, 2012.

project continues to progress, the risk decreases and the costs increase. Figure 3.9 shows the rough breakdown of the risk and cost levels throughout project development.

Phase 1: preliminary survey

The preliminary survey phase consists of identifying geothermal areas on a national or regional scale based upon previous studies and available data. If no previous data is available for analysis, data acquisition needs to be done. This would typically include remote sensing imaging, geological mapping, geothermal mapping, and surveying photography. Not only should the project look at the physical resource itself, but also studies regarding the countries power market, the current supply and demand, the electrical grid infrastructure, environmental laws and restrictions, and basic accessibility to infrastructure (e.g., roads) should be addressed (Gehringer and Loksha, 2012).

In this phase, all aspects regarding the project development, such as land acquisition, utility regulations, land and mineral rights, and legal restrictions for grid access need to be attended to. For large projects, a complete Environmental Impact Assessment (EIA), should be completed. The correct licensing and permitting for exploration need to be obtained as well. Unfortunately, the acquisition of the required information, permitting, and data collection require time and capital. It is estimated that this phase will take up to a year (Gehringer and Loksha, 2012). The high risk of the initial phase can deter investors and create problems acquiring the required capital.

Phase 2: exploration

The exploration phase can be split up into to three exploration types: geological analysis, geochemical sampling, and geophysical sampling. The geological analysis can consist of a variety of different aspects. Surface mapping of geothermal features (Figure 3.10.1), temperature sampling (Figure 3.10.2), sediment analysis, and geological sampling can provide information regarding the geothermal heat source and location.

Once a geothermal area has been identified and mapped, then geochemical samples of the hot springs and fumaroles can be taken (Figure 3.10.3). The geochemical samples will then be analyzed to determine the chemical and mineral elements and their concentrations within the fluids or gases. The chemical analysis can help determine reservoir origins and the degree of permeability within the rock structure. These results can be input into geothermometer equations to give information regarding the reservoir temperature.

For geophysical sampling, there are two main methods used to measure the conductivity and resistivity of the rocks at depth. The Transient Electro Magnetic (TEM) method and the Magneto Telluric (MT) method. These will give information regarding the geological structures, boundaries and possible anomalies in the subsurface. This allows for the identification of possible faults, fractures or other geologic structures at depth to determine the structure of the geothermal reservoir. Figure 3.10.4 shows the main monitoring station for the MT method.

The total cost of the exploration phase is typically budgeted from US$1–2 million, but can vary up to US$3 million (Gehringer and Loksha, 2012). This is dependent on the size of the area being studied, the types of exploration required, and the level of analysis previously completed for the area.

Phase 3: test drilling

From the information gathered in the two previous phases, locations for potential exploratory wells will be selected and a drilling program will be developed to verify the previous results. Dependent on the proposed project and estimated geothermal reservoir size, anywhere from two to five exploratory wells can be drilled.

Standard drill rigs, as seen in Figure 3.11, can be quite large and expensive to operate. Therefore, slim hole wells, which are under 15 cm in diameter rather than the standard 20–25 cm wells, are becoming more common for exploration wells. Some of the advantages to using a slim hole well, is that it requires a smaller drill rig, thus it becomes more economical and reduces the environmental impact. The use of slim hole drilling can increase the difficulty of drilling through harder rock types and present challenges for coring analysis. During the drilling process, various tests can be conducted in the well ranging from temperature sampling, pressure analysis, live camera feeds, and drill cutting analysis. These tests will help determine reservoir conditions and structure.

Figure 3.10.1 Surface mapping of solfataras in the Krýsuvík volcanic field, Iceland.

Source: Photo by author.

Figure 3.10.2 Temperature sampling at Austurengjar in the Krýsuvík volcanic field, Iceland.

Source: Photo by author.

Figure 3.10.3 Geochemical sampling of a fumarole in the Krýsuvík volcanic field, Iceland.

Source: Photo by author.

Figure 3.10.4 Magneto Telluric (MT) sampling equipment.

Source: Photo by author.

Figure 3.11 Standard table top drilling rig, in place at Svartsengi geothermal field, Iceland.

Source: Photo by author.

Typically, a slim hole well can cost anywhere from US$0.5–1.5 million (Gehringer and Loksha, 2012). Depending on if the project continues into Phase 5 and the performance of the exploration wells, these wells may be converted into production wells or could be used for the starting location for directional drilling.

Phase 4: project review and planning

Once Phase 3 is complete, all the results from Phases 1–3 need to be evaluated. A feasibility study will be performed upon these combined results. This study will compose primarily of three sections; the financial breakdown, a drilling program proposal and the engineering designs required for the power plant and infrastructure.

The financial calculations will include the previous costs for Phases 1–3 and all the other costs for every aspect of the projects future, from legal contingencies to environmental issues. The engineering designs will determine what will be built and what infrastructure is required to carry out the drilling program. Once all the plans are developed, approved, and proper permitting acquired, the project may move onto Phase 5.

Phase 5: field development

Phase 5 sees the drilling of production and reinjection wells, as well the construction of the infrastructure to connect the wells to the central power plant (Figure 3.12). The scale of the project and the estimated power production will

Figure 3.12 Geothermal wellhead and piping at Hellisheiði power plant, Iceland.
Source: Photo by author.

determine the number of wells needing to be drilled. Typically, a production well for a large-scale project is expected to produce 5 MW of electrical power (Gehringer and Loksha, 2012).

The amount of time for this phase will vary from project to project, but for a standard drill rig, such as that in Figure 3.11, with typical coring diameter (20 cm) in a typical volcanic environment, it takes about 40–50 days to drill a 2000 meter deep well (Gehringer and Loksha, 2012). The timeline for drilling each well will vary depending on the geological conditions and the proposed drilling plan.

Phase 6: power plant construction

Once the production wells, reinjection wells, and associated substructure is complete, then the construction of the power plant begins. The power plant requires significant infrastructure other than just the power plant with the turbines. A typical geothermal power plant will consist of separators (Figure 3.13), generators, condensers, cooling towers (Figure 3.13), the power plant with its turbines, and an electrical substation to connect to the transmission grid. On a general basis, the construction cost of the power plant will be between US$1–2 million per MW installed at the plant, excluding the cost of the electrical substation and transmission grid connections (Gehringer and Loksha, 2012).

Figure 3.13 The separators and cooling towers for the Hellisheiði power plant, Iceland.
Source: Photo by author.

Phase 7: start-up and commissioning

As the final phase before full operation, Phase 7 involves finding and resolving any issues with the function of the plant and its connection to the grid. The power plant engineers and construction company must ensure the plant runs efficiently and up to the industry standards set out in the original proposals. As issues are not always present immediately, but develop over time, this typically takes 6–12 months to complete (Gehringer and Loksha, 2012).

Phase 8: operation

Now that the power plant is fully functioning and providing electrical power to the grid, the operation and maintenance (O&M) phase begins (Figure 3.14). This phase is split up into the O&M for the production wells, reinjection wells, and associated infrastructure, as well as the O&M for the power plant and its substructures. The O&M for the production and reinjection field consists of ensuring the pipes and well heads are kept clean and scaling issues are dealt with. They also include the drilling of make-up wells, if production capacity lags. It is estimated that this will cost US$1–4 million per year (Gehringer and Loksha, 2012).

For the O&M of the power plant, the chemical composition of the geothermal fluids will highly impact the amount of maintenance required. If the fluid tends to be extremely corrosive and contains chemicals resulting in increased scaling, the maintenance costs will increase. A typical 50 MW operation would expect

Figure 3.14 Nesjavellir power plant, Iceland.
Source: Photo by author.

O&M costs for the power plant to be US$1–4 million per year. Combined with the O&M costs for the wells and pipelines, the total would be US$3.5–10.5 million per year for a 50 MW power plant (Gehringer and Loksha, 2012).

Development of geothermal energy in the Arctic

Arctic geothermal: Iceland as a developmental model

As one of the forerunners in geothermal development, Iceland stands as a potential developmental model for geothermal energy in the Arctic. The island of Iceland, or Ísland in Icelandic, is situated just below the tip of the Arctic Circle. Only a small portion of the island of Grímsey, off the north coast of Iceland, crosses into the Arctic Circle. Depending on how you define the Arctic, Iceland can certainly be considered a part of the Arctic.

Nearly 100 years ago, Icelanders began to utilize geothermal energy for electricity. The power plant Bjarnarflag was made operational in 1969 generating 3 MW of electricity (Landsvirkjun, 2017). By 2016, geothermal provided 24.32 percent of all electricity production in Iceland, roughly 663 MW of electricity (Orkustofnun, 2016b). Not only is geothermal used for electrical production but

Figure 3.15 Fumaroles and solfataras in the Bitra geothermal system, Iceland.

Source: Photo by author.

nine out of ten homes use heat and hot water from geothermal resources (Orkustofnun, 2016a).

Iceland is the largest subaerial section of the mid-Atlantic ridge. With the tectonic rift situated through the island itself, Iceland is the perfect area for volcanic systems and geothermal utilization. Within the island itself there are 20 different volcanic systems, with 30 active volcanos, creating 33 high temperature geothermal systems (Armannsson, 2016).

There are three main companies who produce electricity and direct use from geothermal systems. Orka Náttúrunnar owns three geothermal power plants producing 423 MWe and 1100 MWt. Their largest operation is the Hellisheiði Geothermal Plant with 303 MWe capacity and it is one of the largest geothermal plants in the world (Orka Náttúrunnar, 2014). HS Orka is responsible for two power plants on the Reykjanes Peninsula, totalling 175 MWe and 190 MWt (HS Orka, 2016). Landsvirkjun has two geothermal plants with a total installed electrical capacity of 63 MWe (Landsvirkjun, 2017).

As over 90 percent of the buildings in Iceland are heated using geothermal district heating methods, there is an intensive infrastructure network built around the country. In Reykjavik itself, part of the district heating comes from the Hellisheiði and Nesjavellir power plants. A 20 km piping system connects the power plants to the district heating system in the city. As the fluid moves through at max speeds of 2250 l/s, it only drops about 2°C in temperature (Orka Náttúrunnar, 2014). The pipes are heavily insulated with over 30 cm of specially designed rock wool.

Despite the large amount of geothermal fluid coming from the power plants, it does not meet the demand. Therefore, throughout the city there are a series of low temperature wells drilled between 800–1100 m deep. These wells are all connected to pumping stations which will then mix the water from the wells, which is between 100–120°C, with the cooler recycled water to create a working fluid at 80°C. This fluid is then pumped to the residential and industrial areas for heating and direct use. The city has even implemented a water storage facility, known as the Perlan, in case the district heating system should encounter problems.

As there are still smaller communities and residences in the more remote portions of the country, they are still working on developing access to the districting heating grid. There is a well-developed electrical grid across the country to mitigate the fact that most of the power plants are 20–50+ km away from the main residential and industrial areas.

There is still continuous data being collected and technologies developed in the geothermal industry. Most recently, the Iceland Deep Drilling Project just completed drilling of the IDDP-2 well on the Reykjanes Peninsula. This project took an existing geothermal well drilled to 2500 m and extend the depth by 3000 m. On January 25, 2017, they successfully drilled the well to 4659 m reaching temperatures of 426°C (IDDP, 2017). This opens more options to start researching higher efficiency energy production using super critical fluid. Another option is deep reinjection into geothermal systems.

Through the implementation of renewable energy resources, the National Energy Association of Iceland claims that the Icelandic population has saved US$7.2 billion since 1970 (Orkustofnun, 2016c). Geothermal has helped bolster the economy in Iceland. It has provided not only energy and heating sources, but jobs as well. One example of industry is the business Haustak, an Icelandic company specializing in dried fish. Their fish drying plant, on the Reykjanes Peninsula, uses geothermal heat to run the drying ovens and the electricity used is generated from the Reykjanesvirkjun power plant. Most of the final product is shipped to Nigeria to make fish-head soup, which is a delicacy for their culture.

One new potential aspect to assess is the possibility of off shore geothermal drilling. As the technology for geothermal drilling emulates that of traditional oil and gas drilling, the developmental process has potential for a successful cross-over into offshore drilling. A company, North Tech Energy, recently secured two off shore geothermal exploration permits for two areas off the North and South shore of Iceland. If this proves to be successful, there are possibilities for the further development off the coast of Alaska, Northern Japan and Northern Russia (Richter, 2017).

Other Arctic related geothermal projects

As of 2016, the United States currently has 3.7 GWe of installed geothermal electrical capacity. One geothermal project is of key interest. The Chena geothermal project in Alaska, is the forefront for Arctic geothermal for the United States. This low to moderate temperature hot spring, is one of 30 located in the interior of Alaska (Erkan et al., 2007).

Up to 20 boreholes have been drilled between 100–1000 feet deep (30–304 m). The hottest temperature they have encountered is 176.5°F (80°C), but geochemical results suggest temperatures up to 250°F (121°C). The main purpose of their project is to continue with geothermal exploration to characterize surface geology, borehole geology, reservoir structure, and fluid-rock interactions (Erkan et al., 2007). Although geothermal exploration is still undergoing in this area, there is a 0.25 MW power plant currently operational at Chena Hot Spring, with the potential to expand power generation in the area.

Canada currently does not have any operational geothermal power plants. Progress is being made with the possible construction of a power plant in Valemount, British Colombia by a Canadian geothermal company called Borealis GeoPower. The more promising project regarding developmental status is currently near Estevan, Saskatchewan. In 2017, the Deep Earth Energy Project (DEEP), signed a Power Purchase Agreement (PPA) with SaskPower and has the go ahead to develop a 5 MW power plant (DEEP, 2017).

Future development in the Arctic

As the Arctic is still underdeveloped, in the sense that the majority of it is still untouched by human development, the potential for geothermal energy is unknown. Although geothermal may seem like an ideal solution as a long-term base load energy source, many challenges will be faced. The potential for district heating and direct use of any geothermal system is higher than for electricity development. The ideal conditions, as seen in Iceland, for geothermal power plants, are strictly tied to volcanic systems, thus the use of heat pumps and direct use for domestic purposes is more likely.

There must be a developmental plan in place to assess, regulate, and determine how to progress in this field. Private and public funding opportunities, with incentives, need to be made available to attract interest in geothermal projects. Without the support of the governmental institutions and communities who will benefit from the projects, the industry does not have opportunities to develop. The surface and preliminary geothermal exploration phases are just the beginning of the long process to establish geothermal as a source of power. With the development of projects, the need for proper EIAs and mitigation strategies need to be developed as well. For projects to have the best success, collaboration with other industries should be explored, such as the inclusion of spa benefits or tourism activities, as seen in Iceland. The community support will increase, if more benefits for the community can be produced.

Summary

In summary, the Arctic holds many unknown factors that will determine the feasibility for geothermal. The geothermal industry continues to grow, but its development in the Arctic, outside of Iceland, is very slow coming. Without further support from governments, investors, and the public in general, the developmental process will continue to crawl along. There is a distinct lack of regulatory frameworks in the geothermal industry outside of areas with significant geothermal industries. Iceland is a geothermal industry leader for not just technology and innovation, but also regulations, knowledge, and implementation strategies for designing direct use and electrical systems in Arctic conditions. This knowledge is invaluable and should be built upon by other countries beginning to enter the geothermal sector. The advances seen in development in Iceland have helped to shape the global geothermal industry and the development in the Arctic should be no different.

References

Allis, R. (2000) "Insights on the Formation of Vapor-dominated Geothermal," *Proceedings World Geothermal Congress*. Kyushu – Tohoku, Japan.

Armannsson, H. (2016) "The Fluid Geochemistry of Icelandic High Temperature Geothermal Areas," *Applied Geochemistry*, Vol. 66, pp. 14–64.

Axelsson, G. (2013) "Tracer Tests in Geothermal Resource Management," s.l., *EDP Sciences*, Vol. 50.

Blodgett, L. (2014) "Geothermal Basics," http://geo-energy.org/Basics.aspx, accessed October 2, 2017.

DEEP (2017) "Developing Eothermal Resources to Meet Increasing Energy Needs with Sustainable, Clean and Renewable Energy," www.deepcorp.ca/, accessed November 29, 2017.

DiPippo, R. (2012) "Geothermal Power Plants: Principles, Application, Case Studies and Environmental Impact," Third ed. *Elsevier*, North Dartmouth, Massachusetts.

Eliasson, E. T., Thorhallsson, S., and Steingrímsson, B. (2011) "Geothermal Power Plants," *UNU-GTP*, Santa Tecla, El Salvador.

Erkan, K., Holdman, G., Blackwell, D., and Benoit, W. (2007) "Thermal Characteristics of the Chena Hot Springs Alaska Geothermal System," *Standford University*, Standford, California.

Gehringer, M. and Loksha, V. (2012) "Geothermal Handbook: Planning and Financing Power Generation," *Energy Sector Management Assistance Program (ESMAP)*, Washington DC, United States.

HS Orka (2016) "Power Plants," www.hsorka.is/en/power-plants/, accessed November 28, 2017.

IDDP (2017) "The Drilling of the Iceland Deep Drilling Project Geothermal Well at Reykjanes Has Been Successfully Completed," *Iceland Deep Drilling Project*, Reykjavik, Iceland.

IGA Service GmbH (2014) "Best Practices Guide for Geothermal Exploration," *IGA Service GmbH*, Bochum, Germany.

International Energy Agency (2017) "Oil," www.iea.org/about/faqs/oil/, accessed October 10, 2017.

Kanoglu, M. and Bolatturk, A. (2008) "Performance and Parametric Investigation of a Binary Geothermal Power Plant by Exergy" *Renewable Energy*, Vol. 33, No. 11, pp. 2366–2374.

Kwambai, C. (2014) "High Temperature Geothermal Power Plants and Overview of Wellhead Generators," *Short Course IX on Exploration for Geothermal Resources*, Lake Bogoria and Lake Naivasha, Kenya.

Landsvirkjun (2017) "Power Stations," www.landsvirkjun.com/company/powerstations, accessed November 28, 2017.

Lindal, B. (1973) "Industrial and Other Applications of Geothermal Energy," in H. Armstead (eds) *Geothermal Energy*, UNESCO, Paris, pp. 135–148.

Lund, J. W., Bjelm, L., Bloomquist, G., and Mortensen, A. K. (2008) "Characteristics, Development and Utilization of Geothermal Resources – a Nordic Perspective," *Enterprises*, Vol. 31, No. 1, pp. 140–147.

Lund, J. W. and Boyd, T. L. (2016) "Direct Utilization of Geothermal Energy 2015 Worldwide Review," *Geothermics*, Vol. 60, pp. 66–93.

National Renewable Energy Laboratory (NREL) (2010). "Geothermal Energy Production with Co-produced and Geopressured Resources," *US Department of Energy, Office of Energy Efficiency and Renewable Energy.*

Orka Náttúrunnar (2014) "ON Power About Us," www.onpower.is/about-us, accessed November 28, 2017.

Orkustofnun (2016a) "Direct Use of Geothermal Resources," www.nea.is/geothermal/direct-utilization/, accessed November 28, 2017.

Orkustofnun (2016b) "Installed Electrical Capacity and Electricity Production in Icelandic Power Stations 2016," *Orkustofnun*, Reykjavik, Iceland.

Orkustofnun (2016c) "The National Energy Authority," www.nea.is/the-national-energy-authority/ accessed November 28, 2017.

REN21 (2017) "Renewables 2017 Global Status Report" *"REN21,"* Paris, France.

Richter, A. (2017) "Offshore Geothermal Development – and Option for the Arctic?," www.thinkgeoenergy.com/offshore-geothermal-development-an-option-for-the-arctic/, accessed November 29, 2017.

Tester, J. W., Drake, E. M., Driscoll, M. J., Golay, M. W., and Peters, W. A. (2012) "Sustainable Energy Choosing Among Options," Second Edition, *MIT Press Books*, Massachusetts.

World Energy Council (2016) "World Energy Resources Geothermal 2016," *World Energy Council*, UK.

Wright, M. (1998) "Nature of Geothermal Resources," in J. W. Lund (eds) *Geothermal Direct-Use Engineering and Design Guidebook.* Klamath Fall, OR, Geo-Heat Center, pp. 27–69.

4 Modelling of Longyearbyen's energy system towards 2050

Enzo A. J. Diependaal and Hans-Kristian Ringkjøb

Introduction

This chapter will discuss the challenges of designing a sustainable energy system in the Arctic through the analysis of a working paper written for the University Centre in Svalbard (UNIS).[1] The working paper, 'Modelling of Longyearbyen's Energy System Towards 2050' (Diependaal *et al.*, 2017) identifies the possibilities, challenges and associated costs of a sustainable energy system through the analysis of three different energy scenarios for Longyearbyen's future.

Longyearbyen is Norway's main outpost within the Svalbard archipelago. Svalbard seeks out to be one of the world's best-conserved wildernesses and only a few settlements are found on its main island, Spitsbergen. These settlements are predominantly powered by the burning of fossil fuels, primarily coal and diesel fuel. Longyearbyen has the questionable honour of hosting the last Norwegian operated coal-fired power plant, which is one of the two remaining coal-fired power plants on Spitsbergen. The other power plant is operated in the Russian settlement of Barentsburg.

Ever since John Monro Longyear established Longyearbyen in 1906 for coal mining operations by his 'Arctic Coal Company', the town has been closely linked to the coal industry. The current mining company, Store Norske Spitsbergen Kulkompani (SNSK), took over operations and property in 1916. Since the 1990s, the town has seen a diversification of its economic activities after a combined resource and market crisis left the coal company, and with it the community, vulnerable. Since, diversification has ensured that the town grew to more than 2000 permanent residents, hosting additional assets such as UNIS, Svalbard Satellite Services and the Svalbard International Airport. Current low coal prices create the same scenario for the coal industry and its future is yet again unsure (Misund *et al.*, 2017).

As of 2017, Store Norske operates only one active mine, 'Gruve 7', after the 2016 closure of its Lunckefjell mine in Svea. Closure was inevitable due to declining coal prices and waning political support for coal mining. 'Gruve 7' is left open for the supply of coal for the power plant and excess production is sold and shipped out. In the light of the aging energy infrastructure, the insecurity of the energy supply through overpriced and subsidized coal and the mounting concerns and (political) pressure regarding climate change and a polluting industry in the Arctic, there is a need to evaluate future energy supply options for Longyearbyen.

Past research focused on opportunities for the possible creation of a sustainable energy system for Longyearbyen utilizing the entire coal infrastructure. The energy system would use carbon capture and storage (CCS) to close the local carbon cycle and create a zero-carbon or negative-carbon emissions community. CCS research was initiated in 2007, after a paper by UNIS-director Gunnar Sand and Professor Alvar Braathen, where they propose to turn the town's main polluter into a showcase for CCS using sub surface aquifers (Sand and Braathen, 2006). However, with the current shift towards the integration of alternative or renewable energy sources, research moved away from coal options altogether, reducing interest in the CCS option.

Replacing coal, which has the highest output of greenhouse gases per unit of energy produced, would result in the largest reduction of greenhouse gases. Replacing the system with sustainable or renewable energy sources could illustrate a successful energy system for Arctic communities and provide a showcase for the Norwegian government.

To assess the feasibility and the development potential of a sustainable or renewable energy system in the Arctic, the Integrated MARKAL-EFOM System (TIMES) was used in the 2017 working paper (Diependaal *et al.*, 2017) to build a model of the energy system in Longyearbyen. The working paper focused on three different possible scenarios for the energy future of Longyearbyen:

- a scenario including all technological options, including fossil fuels with coal imports;
- a zero-carbon emissions sustainable scenario including experimental nuclear energy;
- a zero-carbon emissions sustainable scenario with only renewables (Solar PhotoVoltaic (PV) and wind power).

The Integrated MARKAL-EFOM System

TIMES, a model generator software package developed by IEA-ETSAP (International Energy Agency-Energy Technology Systems Analysis Program), provides a technical engineering approach and an economic approach for long-term energy scenario analysis for energy systems on a local, national, multi-regional or global level. It uses a technology-rich basis for representing energy dynamics over a multi-period time horizon. The estimated end-use of energy drives the reference scenario. The user supplies the stock of the available energy sources and conversion methods, together with possible energy and environmental criteria (Loulou *et al.*, 2016).

The TIMES model aims to supply the energy at a minimum global cost by simultaneously making decisions on equipment investment and operation; primary energy supply; and possible trade. This makes TIMES a vertically integrated model for the entire energy system. The core model utilizes the dynamic partial equilibrium for supply and demand with perfect foresight. Within TIMES, this perfect foresight assumption extends towards the model horizon (Loulou *et al.*, 2016).

As the model uses physical, environmental and political constraints, it can be used to model the long-term effects of energy-environmental policy and legislation. In the case of the Longyearbyen model, it is used as a cost estimate model for the introduction of a fully sustainable energy future through the analysis of the three scenarios as compared to the current system.

Demand

TIMES uses the projected energy demand towards the model horizon as one of its major drivers. It is therefore a highly important input variable defining the validity of the entire model outcome. The user must supply this information up front.

To ensure a well-defined estimate for future demand, recent historical demand and future development policies are to be considered. Presently, Longyearbyen hosts 2140 permanent residents. Population development in recent history has seen the population increase approximately 25 per cent since the year 2000. However, current Norwegian policy has expressed the desire to curb this growth and keep the population at current levels.

With the increase in population, a rise in energy demand is expected as heated area and electricity consumption are expected to rise accordingly. For the heat demand, this expected rise seems to hold true. Electricity consumption however declined in the same period (Figure 4.1). The change and differentiation in local industrial activities within the same period is expected to be at the foundation for this discrepancy.

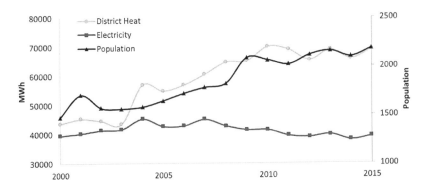

Figure 4.1 Energy consumption and population in Longyearbyen from 2000 until 2015.

Source: Diependaal *et al.*, 2017 (used with permission).

Present-day activities added during the diversification of the local economy do not seem to weigh heavy on the energy demand. The main contributor added to the local economy is tourism, with the accompanying birthing of large cruise ships. While hotels will require additional energy, the ships are self-powered, requiring no external power besides the additional energy use of harbour equipment. The tourism industry is however expected to grow by a considerable amount, possibly tripling total numbers, adding to the heat and electrical demand. Offsetting this predicted growth might arise from a replacement of the building stock with more energy efficient buildings and technologies.

With 20 per cent of current day electricity production being used in local heavy industry, namely the power plant itself and the coal mine, it is reasonable to assume that the reduction in mining activities stands at the base of the electricity decline. Mining declined since peak production hit 4.1 million tonnes in 2007, dropping to 1.1 million tonnes in 2015. This reduction seems to offset the increased demand due to the growth in population.

Closure of the mine and the power plant, by changing to a sustainable solution, would result in a 20 per cent reduction in electric energy demand. However, with a constant permanent population and a rising tourism branch, it is assumed that power and heat usage in the future is constant from present-day levels.

Load profiles

To ensure usage of representable load profiles, an hourly resolution within each period was chosen for this working paper. Each year within the model is represented by one-hour slices over two days, for every season, for a total of 192 slices per year. The decision for two days was made due to the different load profiles during the week and the weekend. Winter is set as the months of December, January and February, spring as March April and May, summer as June, July and August and fall as September, October and November.

Every day was given a load profile for both thermal (Figure 4.2) and electrical (Figure 4.3) demand, to represent the energy usage during that period of the year. These load profiles are calculated using time-slices over multiple days within the season, to provide an applicable average. An important observation is the change in environment during the spring and fall seasons, in which the transition from the polar night to the polar day, and vice versa, takes place. This transition is simplified within the season average.

Two large changes in consumption immediately take notice. There is an extremely large difference between the summer and the winter months in heat and power consumption, as the polar winter requires a large demand for both increased heating and the necessity of artificial lightning. However, the difference between the winter, and both fall and spring is less profound.

Also noticeable is the diurnal cycle, with low electricity consumption during the night and high consumption during the day. Noticeable is the absence of a 'lunch-break dip' within the weekend periods, possibly due to the absence of mining activities. Heat profiles vary much less between day and night, as

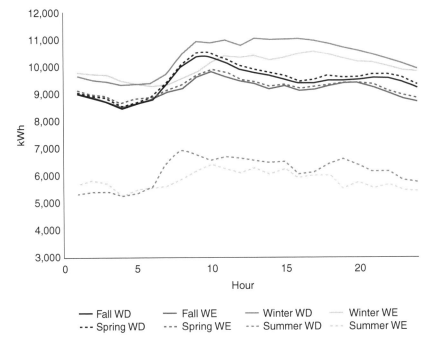

Figure 4.2 Daily seasonal thermal load profiles, for both week days (WD) and weekends (WE).

Source: Diependaal *et al.*, 2017 (used with permission).

temperature during the polar day and polar night remains almost constant over a 24-hour period.

Scenarios

The main objective of the model was to find a reliable energy system with the lowest discounted system cost possible. To vary the outcomes of the model, three scenarios were used:

*   A scenario including all options, by continuation of the present-day system through coal imports.
*   A zero-carbon emissions sustainable scenario including experimental nuclear energy.
*   A zero-carbon emissions sustainable scenario with only renewable energy sources (Solar PV and wind power).

The present-day system was modelled and calibrated using data from the power plant. Solar PV and wind power were analysed for their respective challenges,

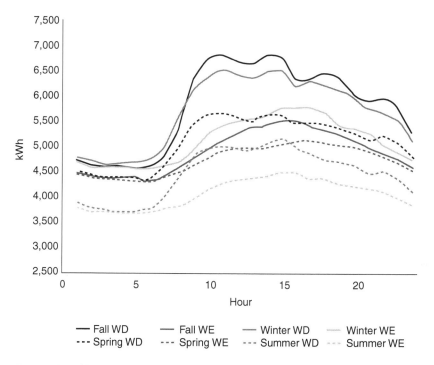

Figure 4.3 Daily seasonal electrical load profiles, for both week days (WD) and weekends (WE).

Source: Diependaal *et al.*, 2017 (used with permission).

cost drivers and production capabilities. Finally, the preliminary data from an
experimental nuclear Small Modular Reactor (SMR) was used to determine
operational parameters and total lifetime expenditures.

Present-day system

The current energy system is built up around a central coal-fired power plant,
which supplies both electricity and heat for the residents of Longyearbyen. Coal
is supplied by 'Gruve 7' and consumption is approximately 25,000 tons per year.
Built in 1982, the power plant faces challenges of ageing equipment; however,
current upgrades ensure an operational life until 2038. In contrast, current esti-
mations place the available mineable coal reserves in 'Gruve 7' between seven
and ten years. After that, coal will have to be extracted at another claim, or
imported. Alternative fuels, such as wood pellets or natural gas, are considered
(but not included within this study).

The current power plant has two boilers, supplying steam to a back-pressure
(non-condensing) turbine and a condensation turbine. The back-pressure turbine
acts as the base load turbine, supplying both heat and power to the city. The

condensation turbine is used as the load following turbine, following the grid demand during the day.

Backup systems are in place to provide energy security. The power plant itself has a heat exchanger that can directly convert steam from one of the two boilers into additional heat capacity for the city district heating system, in case the back-pressure turbine is out of operation, further increasing the energy security. Besides the internal back-up system, the town uses five diesel generators as backup generation capacity. Six fuel oil fired boilers are placed strategically over town to provide backup for the city district heating system and to act during peak loads. Strategic placement is necessary in case a certain section of the infrastructure fails, to provide segmented backup capabilities.

Solar PV

Solar energy is one of the considered renewable energy sources in the Arctic, through means of PV cells. While the polar winter offers no solar energy, the polar day provides 24 hours of sunlight, however at lower intensity as compared to lower latitudes. This reduction in intensity is slightly offset by an increased efficiency due to lower mean operating temperatures. This makes solar a sensible option during the summer, but one that requires extensive battery power if being the only considered energy source.

PVSyst was used for initial analysis of the expected solar energy. PVSyst is a comprehensive software package that allows for the modelling of PV systems in both off-grid and grid-connected systems. To calculate the output of a certain PV system, meteorological data from Meteonorm is used. Meteonorm calculates the local average ambient temperature and global horizontal irradiation data by interpolating the values from (usually) the three nearest weather stations.

For the TIMES model, a generic value for energy production is needed, as the model itself will scale the PV system accordingly to create the least cost solution. In this case, two types of panels with different efficiencies and prices were considered. These were placed under 30° in a southern direction (optimal placement on a ground placed structure) or under 20° in a southeast and northwestern direction (as the local roof tilt would dictate).

Analysis showed that the more costly, high efficiency monocrystalline panel only a has slightly higher capacity factor (due to increased production at low light conditions, barely visible as the dotted lines in Figure 4.4), not justifying its price. Furthermore, the total capacity factors per season are obtained, reaching a maximum of 0.4 on a south-facing panel during the spring. During the summer, the capacity factor is slightly lower. Assumptions are that this phenomenon arises from increased overcast, fog and temperatures during this season.

Wind power

Wind power has its own challenges within the Arctic region. These challenges can be categorized in environmental and geographical threats. The environmental

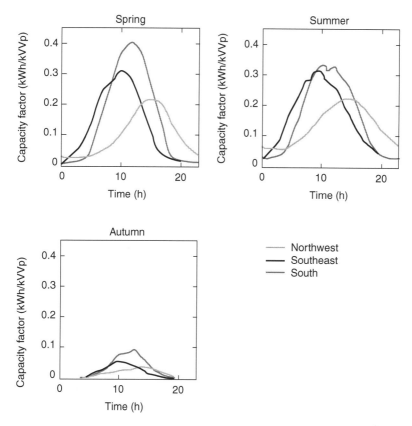

Figure 4.4 Capacity factor comparing different panel direction in the different seasons (with winter being zero), retrieved from PVsyst. For autumn and spring, average daily irradiation is used, as these seasons see very rapid changes from full light to full darkness. Dotted line, almost completely overlain, gives the slightly increased capacity factor for the monocrystalline panels.

Source: Diependaal *et al.*, 2017 (used with permission).

challenges originate from the harsh weather conditions that are often encountered at these high latitudes. These weather conditions present themselves in different sorts of precipitation, large temperature ranges and varying wind conditions. Geographical challenges are imposed by the specific conditions given by the surrounding landscape. An open tundra influences a wind turbine vastly different than mountainous terrain. Both features vary greatly throughout the Arctic and should be re-evaluated for every different project.

On Svalbard, the weather is extreme and can change quickly. The first obvious variable is the wind itself and meteorological data indicates that the quality of the wind resource is rather poor. Average yearly wind speeds are, depending on the specific location, around the 5 m/s mark. Monthly averages provide the highest average wind speed during the winter months, averaging approximately 6 m/s (Petersson, 2007). This would qualify even the best months as being a poor resource, possibly resulting in low capacity factors if the turbines are not optimized for these speeds. Furthermore, the wind is highly variable and dependent on local factors. Gusts as high as 41.2 m/s have been recorded (Sjöblom, 2014), adding further to the technical challenges. High velocity gusts can cause a wind turbine to cut out to prevent structural damage, reducing the capacity factor even more.

The second important variable is temperature. Temperatures lie below freezing for extended periods of time during the year. These low temperatures have their effect on the materials used in the wind turbine design. One should account for the effects of cold climate operations as material properties might change considerably. Examples include materials that become brittle and lubricants becoming waxy, creating different operation parameters and possibly inflicting a shorter time between failure or a reduced life expectancy for the turbine. Other direct influences of these low temperatures are often a problem in the form of the turbine blades icing up. Icing describes the formation of ice on the turbine blades and the nacelle, through the freezing of moisture in the air or the collection of precipitation under freezing conditions.

Icing creates different problems, which can threaten both operations and human life. Changes in the aerofoil due to ice reduce the efficiency of the blade, meaning less energy can be extracted from the passing wind. Furthermore, an alteration in aerofoil or induced roughness due to the ice can disrupt the flow of air over the blades and create an unstable boundary layer. These turbulent flows can induce sound, vibrations and varying loads over the blades. Where noise is a mere inconvenience for residents and nature in the direct vicinity of the turbine, vibrations and varying loads can cause an increased wear on the turbine or structural damage. A more life-threatening situation arises when ice starts to break off the blades. This phenomenon can create high mass, high velocity projectiles, which can cause harm and threaten both life and infrastructure over large distances.

Topography dictates the local condition to a large extent. Svalbard features a landscape dominated by fjords. These fjords create different wind conditions as compared to more open landscapes and have similarities with other mountainous terrains. The prevailing wind direction is heavily dependent on the local topography, as the fjord's walls can deflect geostrophic wind through the process of forced channelling. This process can amplify the velocity of the wind, creating local wind speeds far higher than the uninfluenced average.

Another side effect arises when multiple winds are channelled together and (re)combine. While channelling of wind can create higher wind speeds, increasing the amount of energy in the wind, it usually creates heavy turbulence when winds of different speeds combine again. This turbulent flow affects a turbine in a negative way by causing disturbance of the boundary layer, resulting in reduced efficiency and induced fatigue. Wind turbine placement should account for these local features, to reach more favourable wind conditions.

Wind power in Arctic regions should favour designs that take the extreme environment into account, while accounting for additional costs for operations and maintenance (O&M). Given the lack of infrastructure in most Arctic regions, the placement of wind turbines in the Arctic can add considerably to the total cost. Placement in an undisturbed flow, out at sea or high on a mountain, can make the total cost rise more quickly than in the same conditions at lower latitudes.

Small modular reactor

Renewable energy sources are intermittent and therefore an abundant amount of backup must be created to ensure continuous energy availability, increasing the total system cost. For the second scenario, the unproven technology of a (very) small modular reactor (vSMR) was chosen to test the case for a dispatchable sustainable energy source. While Longyearbyen has multiple options for other dispatchable energy sources (local resources of natural gas are discovered and the import of fuels such as wood pellets is considered an option), it would mostly use the existing infrastructure through fuel combustion. The new system would replace the current power plant and therefore provides a more interesting case.

Norway has stated in the past that nuclear energy would not be a part of its energy mix. Until now, cheap hydro can deliver 99 per cent of all electricity produced in Norway (Statkraft AS. 2009) and therefore, nuclear energy is not necessary. However, with the recent development and increased international interest in small modular reactors (most of them between 10 and 300 MW), the use of nuclear energy for small isolated grids becomes interesting.

Currently, Norway has no functioning commercial nuclear power stations, or the intention to build one. However, Norway does have a framework regarding the licensing and construction of nuclear facilities as it hosts two research reactors (OECD, 2001).

While Norway placed a moratorium on nuclear energy 35 years ago, it does not oppose the use of nuclear energy. Besides its legal framework, the Thorium Report Committee, on behalf of the Ministry of Petroleum and Energy, announced in 2008 that Norway should increase its research in the nuclear industry, focusing on thorium-fuelled reactors, due to extensive domestic resources. The Thorium Report Committee states that: 'No technology should be idolized or demonized. All carbon dioxide (CO_2) emission-free energy production technologies should be considered. The potential contribution of nuclear energy to a sustainable energy future should be recognized' (Kara *et al.*, 2008). These facts and developments lay the foundation for nuclear technology assessment in this report.

For the isolated grid of Longyearbyen, a vSMR was assessed for its suitability. Given the energy use of the current system, the U-battery, developed by URENCO in partnership with Amec Foster Wheeler, Cammell Laird and Laing O'Rourke, proved applicable, due to its small size and extensive documentation (URENCO, 2017). The U-battery is also assessed for the utilization of thorium as nuclear fuel. The development of thorium as a nuclear fuel makes the U-battery an even more interesting candidate for this project, given the advice on future nuclear research in Norway (Ding *et al.*, 2011).

U-battery is build-up of small nuclear cores, which can supply a combination of heat and power to the local grid. The power output of each core is 10 MWth (thermal output), and can deliver up to 4 MWe (electrical output). The system is designed to be put in place with two cores, delivering 20 MWth and 8 MWe and possesses load following capabilities. Load following allows the power plant change its output according to the heat and electricity demand at each time, previously impossible and/or unwanted with older generation nuclear power plants.

Cost estimates

TIMES requires the specific investment and O&M cost for each technology for its decision making. Costs for wind, solar, heat pump and electric boiler were retrieved from the Norwegian Water Resources and Energy Directorate (Norges vassdrags- og energidirektorat, 2015). The costs for both wind and solar have been increased by 20 per cent from their published value to approximate the influence of the harsh Arctic environment and additional transportation. To account for current developments in these technologies, it was assumed that the specific investment cost will drop until the model horizon in 2050. Heat pump and electric boiler technology are assumed to have constant investment cost within this model.

The assumed cost of the vSMR is approached in the 2011 report 'Design of a U-Battery' (Ding *et al.*, 2011). Based on the Cost Estimating Guidelines for Generation IV Nuclear Energy Systems they find that construction costs decline at the rate equivalent of 0.94. This value is used to define the price of an eighth of a kind power plant, providing a more reasonable cost estimation than the first of a kind of the series. This is important as costs of modular reactors are to go down when production progresses, providing for a more realistic economic evaluation. The investment cost for the small modular reactor is also assumed to be constant.

The cost and heating value of the coal was given by power plant manager Kim Rune Røkenes, and equalled 565 NOK/ton and with a heating value of 7.34 MWh/ton. Local engineer for the community council Rasmus Bøckman provided the cost and heating value of diesel. Total cost was 9.25 NOK/litre on with a heating value of 9.5 kWh per litre. An overview of all specific investments cost and O&M cost assumptions is presented within Table 4.1.

Conclusion

If the user permits TIMES to utilize all available energy technologies, the fossil fuelled system comes out as the option with the least expenditures until the model horizon of 2050. The system lacks major investments in new infrastructure and only requires O&M and fuel costs. Within this scenario, it was assumed

Table 4.1 Specific investment and O&M costs assumptions per technology, in MNOK/ MW

Technology	Specific investment cost (MNOK/MW)	Fixed O&M costs (MNOK/MW)
Wind	12.3–54	0.48–1.5
Solar	12.0–21	0.09
vSMR	81.5	0.63
Heat Pump	10.25	0.04
Electric Boiler	0.718	0.004
Battery Storage	2.5	0.151

Source: (Diependaal *et al.*, 2017) (used with permission).

that the price of coal as currently paid by the operator stayed the same over the lifetime of the model. This is however a conservative choice, as coal on the world market is less expensive than the coal that is locally mined. Further assumptions were that the coal-fired power plant could be operated till 2050 without any large upgrades or overhaul necessary.

The model does invest in 0.45 MW of heat pumps and 1.36 MW of onshore wind turbines, for replacement of the current diesel generators and oil boilers. This specific investment happens already in 2020, the first possible decision step. It proves cheaper to invest in two new technologies with low operating costs rather than buying expensive diesel fuel. This change would result in a CO_2 emissions reduction by about 5000 tonnes per year (3.6 per cent of total energy-related emissions). This is a result of the data supplied to the model. In this case, usage of average values and relative large time-slices smooths the production curve of wind power, neglecting extremes. More detail in the data is necessary to further develop this renewable backup system.

In the zero-carbon emission scenarios with all technologies available, TIMES picks the vSMR as its next best option. Despite large investment costs, the vSMR proves to be an attractive option because of its low O&M costs, 60-year long lifespan, high capacity factor, flexibility and the availability to provide both heat and power. One should, however, consider that these results are obtained using preliminary data from experimental technology in its development phase. Data available at more advanced stages would lead to more robust results. To cover for the gaps in the capacity factor, the model invests in 2 MW of onshore wind and 700 kW of electric boilers, combined with an extra 11 MWh of battery storage. The high capacity factor results for the system to require less installed capacity as compared to the current system, as seen in Figure 4.5.

If the model is forced to only use renewable technologies, a high installed capacity of both solar and wind is used to cover total demand. This is a necessity to cover for the relative low capacity factor of both technologies. A further increase arises from the additional electricity demand for electrical heating. The model invests in 26 MW of ground-mounted PV panels, 32.5 MW of onshore wind, 13.8 MW of heat pumps and 5 MW of electric boilers. The installed capacities of

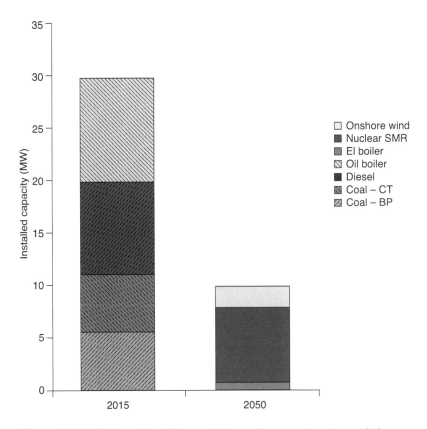

Figure 4.5 Installed capacity difference between the current system and the proposed nuclear-powered scenario.

Source: Diependaal *et al.*, 2017 (used with permission).

solar and wind corresponds to installing respectively 95,000 PV panels and 11 wind turbines. To cover for moments of low production, the investment in battery storage is increased to a total storage capacity of 50 MWh.

The combination of both solar and wind makes sense, as solar energy output is at its peak during the polar day, while wind power production peak during the polar night. This smooths out the production differences between the two technologies. A visualization of production and demand during the diurnal cycle of each modelled day is shown in Figure 4.7.

Preliminary results in the form of total system costs and cost of energy can be found in Table 4.2. The table shows that the most cost-effective option is the continuation of the current system, with import of coal at world market prices. In a zero-carbon emissions scenario, experimental vSMR technology comes out as the best option, with an increased system cost of 38.5 per cent as compared to

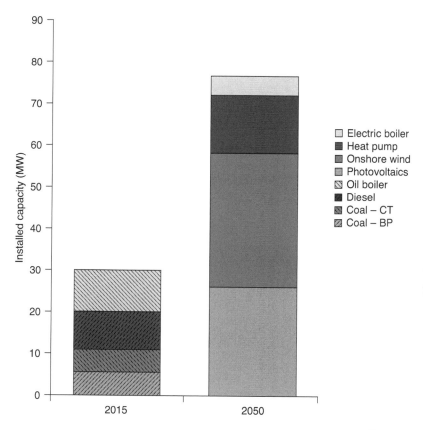

Figure 4.6 Installed capacity difference between the current system and the proposed renewable-powered scenario.

Source: Diependaal *et al.*, 2017 (used with permission).

Table 4.2 Overview of model outcomes, in total expected system cost and total cost of energy

Description	Total system cost	Cost of energy
Existing system until 2050	691 MNOK	1.4 NOK/kWh
Zero CO² in 2050, all techs included	956 MNOK	1.7 NOK/kWh
Zero CO² in 2050, only renewables	1,263 MNOK	2.1 NOK/kWh

Source: (Diependaal *et al.*, 2017) (used with permission).

the current system. However, due to low O&M costs, levellized cost of electricity only increases by 21.4 per cent. Finally, if only renewables are considered, the total system cost would increase by 82.8 per cent as compared to the current system and the levellized cost of electricity would increase by 50 per cent.

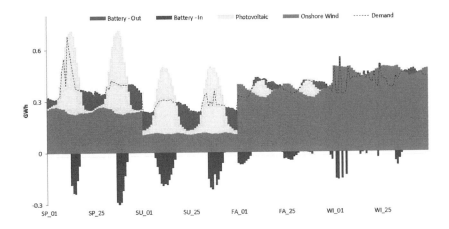

Figure 4.7 Visualization of energy production and demand for each modelled day, including storage.

Source: Diependaal *et al.*, 2017 (used with permission).

The outcome proves that it is possible to provide Longyearbyen with a sustainable future, at an additional cost over the current system. Further research and high resolution data is needed to come to more accurate final answers with regards to each specific sustainable technology, but the data underlines that political support, not technology, is the limiting factor. If the government is willing to commit and provide the right incentives, eliminating coal price volatility and emissions in the process, Longyearbyen could become a sustainable showcase for the Arctic and a sustainable and energy secure community by 2050.

Note

1   This working paper was produced as a group effort during a summer course on Sustainable Arctic Energy Exploration and Development at UNIS. Ringkjøb followed the UNIS PhD course AGF 853, led the group work and performed the modelling while Diependaal, Neumann and Sobrino followed the UNIS master course AGF 353 and contributed to the data collection and analysis. The authors are grateful to UNIS for allowing them the opportunity to address this topic as part of their study programmes.

References

Diependaal, E., Neumann, J., Ringkjøb, H.-K. and Sobrino, T. (2017) *Modelling of Longyearbyen's Energy System Towards 2050.* UNIS, Longyearbyen.
Ding, M., Kloosterman, J. L., Kooijman, T., Linssen, R., Abram, T., Marsden, B. and Wickham, T. (2011) *Design of a U-Battery.* Delft University of Technology, Delft.

Kara, M., Kullander, S., Röhrich, D., Dujardin, T., Kadi, Y., Hval, S., Salbu, B. and Ingebretsen, F. (2008) *Thorium as an Energy Source – Opportunities for Norway.* Thorium Report Committee, Oslo.

Loulou, R., Goldstein, G., Kanudia, A., Lettila, A. and Remme, U. (2016) *Documentation for the TIMES Model,* www.iea-etsap.org/docs/Documentation_for_the_TIMES_ Model-Part-I_July-2016.pdf, [Accessed 25 June 2017].

Misund, O. A., Aksnes, D. W., Christiansen, H. H. and Arlov, T. B. (2017) *A Norwegian Pillar in Svalbard: The Development of the University Centre in Svalbard, UNIS.* Cambridge University Press, Cambridge.

Norges vassdrags- og energidirektorat, (2015) *Kostnader i energisektoren – Kraft, varme og effektivisering.* Norges vassdrags- og energidirektorat, Oslo.

OECD (2001) *Nuclear Legislation in OECD and NEA Countries, Regulatory and Institutional Framework for Nuclear Activities, Norway.* OECD, Paris.

Petersson, C. (2007) *An Analysis of the Local Weather Around Longyearbyen and an Instrumental Comparison.* UNIS, Longyearbyen.

Sand, G. and Braathen, A. (2006) *CO-fritt Svalbard i 2025?* Svalbardposten, 10 November, Longyearbyen.

Sjöblom, A. (2014) *Weather Conditions on Svalbard.* Presentation from the Department of Arctic Geophysics. UNIS, Longyearbyen.

Statkraft AS. (2009) *Hydropower* [online]. Statkraft, Oslo. Available at: www.statkraft. com/globalassets/old-contains-the-old-folder-structure/documents/hydropower-09-eng_tcm9-4572.pdf [Accessed 2 July 2017].

URENCO. (2017) *Prospectus September 2017* [online]. URENCO, Virginia. Available at: www.u-battery.com/_/uploads/U-battery_Prospectus-Sept17.pdf [Accessed 5 January 2018].

Part III

Integrating energy in the Arctic

Infrastructure and communities

5 Towards sustainable energy systems through smart grids in the Arctic

Gisele M. Arruda and Feb M. Arruda

Introduction

Challenges exist in Arctic fossil fuels exploitation ventures and are generally linked to technical, environmental, economic, commercial and social issues. Remote locations with sensitive environments, problems with icing on equipment, safety design of technology, expensive operations and impacts on Arctic communities are a few examples of the twentieth century conventional energy challenges. However the eight Arctic nations have an invaluable opportunity to make genuine and effective changes in how the Arctic is approached to help create a new development and energy models for the twenty-first century Arctic.

The prospect for future energy integration is critical for the progress of a low carbon economy in the Arctic. However, it depends not only on systems integration, but also on the clear integration of policies, investment and technology transfer. In this chapter, the dynamics of innovation and technology for the twenty-first century Arctic energy system and societies will be discussed. By acknowledging the opportunities and challenges of energy integration under the lenses of smart grids, the path towards a more sustainable energy system for the Arctic region becomes more plausible. Other implications and perspectives of the specific dynamics in this region will be mentioned.

Green energy, innovation and technology for the twenty-first century Arctic

Climate change is rapidly transforming the Arctic environment at a speed that will affect local and global societies before they can reasonably understand the physical, social and environmental transformation on course.

Climatic impacts refer to more than just temperature and precipitation. It includes extreme events, such as snow, ice and circulation patterns in the atmosphere and oceans. In the Arctic, sea ice is one of the most important climatic variables. It is a key indicator and agent of climate change that affects surface reflectivity, cloudiness, humidity, exchanges of heat and moisture, the ocean surface, and ocean currents.

According to the latest data from 25 December 2017 provided by the National Snow and Data Center (NSDC), the Arctic sea ice extent for November 2017 averaged 9.46 million square kilometres (3.65 million square miles), representing the third lowest area in the 1979 to 2017 satellite record. The data provided revealed that:

This was 1.24 million square kilometers (479,000 square miles) below the 1981 to 2010 average and 830,000 square kilometers (321,000 square miles) above the record low November extent recorded in 2016. Extent at the end of the month was below average over the Atlantic side of the Arctic, primarily in the Barents and Kara Seas, slightly above average in western Hudson Bay, but far below average in the Chukchi Sea. This continues a pattern of below-average extent in this region that has persisted for the last year.

(National Snow and Data Center, 2017a)

For the purpose of debating the massive changes referred to in this chapter the sea ice retreat parameter, which has enormous chain reaction effects on the environmental systems in several parts of the Arctic, is considered as a whole. This parameter influences a multitude of factors like ecosystems, glaciers melting rates, indexes of albedo and reflectivity, permafrost exposure, ocean levels, food chain and, consequently, local Arctic communities in a severe and unpredictable way.

Figure 5.2 below shows important change in patterns of the daily sea ice extent for the five previous years before 3 December 2017. According to the data analysed by the scientists of the National Snow and Ice Data of University of Colorado Boulder, the observed September sea ice extent fell close the long-term trend line showing the same pattern observed last year. The forecast is 264,000 square kilometres below the previously observed value measured in 300,000 million square kilometres (National Snow and Data Center, 2017b). This trend is confirmed by Figure 5.3 showing the average monthly Arctic sea ice extent from 1978 to 2017 pointing out a 5.14 per cent decline per decade.

The National Snow and Ice Data Center graph of Figure 5.3 shows that the linear rate of sea ice decline for November is 55,000 square kilometres (21,200 square miles) per year, or 5.14 per cent per decade. The average monthly Arctic sea ice extent from November 1978 until 2017 shows a dramatic decrease in the sea ice extension.

The Arctic system, with its components, plays a key role in the balance of the global climate. As the ice retreats, the current challenges concerning energy production in relation to oil and gas activity will intensify even more the effects of climate change on Arctic ecosystems and communities.

The collaborative effort of using scientific ice prediction data and traditional knowledge can reduce the levels of vulnerability at the same time it strengthens Arctic communities resilience and support-capacity to implement the necessary adaptive strategies in response to climate change. The use of traditional knowledge

Figure 5.1 Sea ice extent, December 2017.

Source: National Snow and Data Center, University of Colorado Boulder, (used with permission).

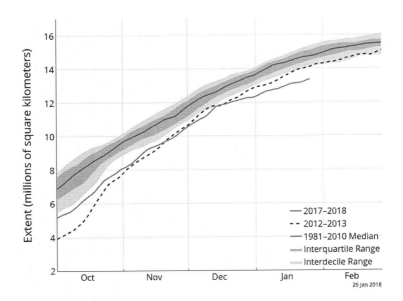

Figure 5.2 Arctic sea ice extent according to sea ice index data (area of ocean with the lowest 15 per cent of sea ice).

Source: National Snow and Ice Data Center, University of Colorado, Boulder.

terminology is an important adaptive tool in terms of the systematization of sea ice science prediction assessment and analysis.

Oil and gas development activities across the Arctic are centenary and responsible for many physical transformations of the 'tangible, environmental place' (Cunsolo *et al.* 2012). The availability of fossil energy resources, because of the ice retreat, has reawakened government interest in the remote parts of the Arctic that can intensify even more the climate change risks associated to conventional energy exploitation.

The problem with fossil fuel exploitation is that large-scale oil and gas development projects do not bring direct economic benefits to local communities from where the resources are extracted. Any significant increases in employment and population are generally short term and offset by erosion in the cohesion of indigenous communities (Arruda and Krutkowski, 2017). Besides this, it increases the amount of Greenhouse Gases (GHG) emissions that countries around the world are trying to reduce under the auspices of the Paris Agreement. The latest GHG emissions assessments are based on the most up-to-date findings from remote sensing, satellite technology employed in climatology data collection and analysis (Roy and Behera, 2017). The integration of different types of remote sensing data, along with ancillary data from old ice cores, makes it is possible to

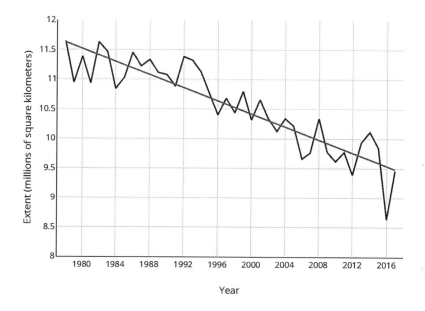

Year

Figure 5.3 Monthly November ice extent for 1979 to 2017 shows a decline of 5.14 per cent per decade.

Source: National Snow and Ice Data Center (used with permission).

cross-analyse data, through the ice layers information, to unfold earlier times in the Earth's climate history and make comparisons to the latest geo-biophysical data available (Rao, 1991).

Remote Sensing (RS) became a crucial technological tool to monitor climate change and its interactions. It is defined by Roy and Behera as:

> the science of identification of earth surface features and estimation of their geo-biophysical properties using electromagnetic radiation as a medium of interaction. Spectral, spatial, temporal and polarization signatures are major characteristics of the sensor/target, which facilitate target discrimination. Earth surface data as seen by the sensors in different wavelengths (reflected, scattered and/or emitted) is radiometrically and geometrically corrected before extraction of spectral information.
>
> (Roy and Behera, 2000)

RS data from European Space Agency's Copernicus and Sentinel – 2 and 3, for example, is also helping to develop a better geographical information system (GIS) thus driving important scientific investigations with applications on climatology, environmental and social sciences. This is done by assessing levels of development, land management, natural resources management, forest biomass, mapping minerals and environmental changes (ESA, 2017). The US Geological Survey (USGS), through the Committee on Earth Observation Satellites (CEOS) provides Climate Data Records (CDRs) that are satellite-based multi-decadal climate data covering terrestrial, oceanic and atmospheric domains at regional and global scales. The CDRs analysis is possible by addressing Essential Climate Variables (ECVs) involving parameters to measure snow cover, glaciers and ice caps; permafrost; land surface albedo; land cover, changes in biodiversity and human concentration rates. Observational data covers changes to the Earth's terrestrial and oceanic environment resulting from natural processes and human activities unveiling consistent monitoring records of climate change effects (USGS, 2017).

This complex Arctic context made urgent the need to decarbonize electricity systems triggering a significant change in how the world produces and uses energy. This change was only possible due to innovation and the evolving technologies related to environmental monitoring, systems of information, communication, energy generation and transmission (Durbha *et al.*, 2017).

The prospects of climate change in very sensitive areas like the Arctic, the increase in fossil fuels prices, the exponential global energy demand and the insecurity of nuclear power are important drivers provoking a fundamental change in twenty-first century energy systems.

This trend can be translated by the expression 'modern grid'. Different from the traditional grid that is centralized, and fossil fuel based, the grid of the twenty-first century is based on transformational technologies connecting a range of different systems consisting in modern engineering, information technology, integrated distribution and transmission in order to provide access, reliability, security and efficiency.

Modern grids or 'smart grids' are considered a key path towards a sustainable energy system with the capacity to reshape economies and societies by transforming geographical energy landscapes. This is relatively clear by understanding the own concept of smart grid as a different energy system from the twentieth century in terms of its decentralization and integration as it has the capacity to embark a wider range of renewable energy sources reducing transmissions losses, infrastructure costs and enhancing energy efficiency (NETL, 2007, 2017).

Smart grids can be conceptualized differently depending on the geographical area, but the definitions have the same core elements when meaning a modern unconventional and integrated grid that uses advanced information technology to update and modernize existing power grids to improve reliability, security, and efficiency of electricity supply systems (Electricity Advisory Committee, 2008).

For the European Technology Platform, a smart grid is an electric network that can intelligently integrate the actions of all users connected to it – generators, consumers and those that do both – in order to efficiently deliver sustainable, economic and secure electricity supplies (ETPSG, 2006). This concept brings on board an important innovative mind-set as, according to it, stakeholders interact with the grid by self-managing the energy system. This is a particularly innovative concept that goes beyond engineering technological advancements as it builds a fundamental bridge with the social sciences.

Last but not least, the core components of this concept provide us with the essential elements to understand smart grids' central role in a twenty-first century energy transition process towards a green-energy technological low carbon economy and society.

Smart grids and energy integration: opportunities and challenges for the Arctic

Travelling across the Arctic presents us with many 'Arctics'. The diverse geographical conditions of different energy potential areas create sites where conventional energy extraction is feasible, other areas where energy extraction is seen as a potential and, in a number of Arctic remote locations, energy is impossible to be extracted according to conventional fossil fuel technologies.

The challenges associated to conventional energy exploration and use are a key driver to trigger renewable energy projects due to the fact that the Arctic environment is not uniform. A new perspective on energy generation has started the process of addressing the energy needs in areas of immediately viable resources recovery, in areas of potential resources extraction and in more isolated zones of harsh environments where resources are unlikely to be extracted even considering all the current available technology in place. This is the reason for saying, commonly, that there is not only one Arctic but many 'Arctics', and the great challenge is how to manage these different 'Arctics' in terms of renewable energy resources, energy policies and energy access to communities. We believe that renewable energy in its different modalities (solar, wind, biomass, geothermal, tidal, etc.), can address the diversity of the Arctic energy landscape by introducing a different perspective oriented to durability and efficiency, but it is still the beginning of a long path.

Additionally, environmental conditions, geological potential and accessibility, population rates, levels of economic development and political leadership are important variants that determine different risk levels in different Arctic territories, as well as the success rates for innovative renewable energy projects.

The balance of risk and opportunity for different models of local and regional energy and human development is a very difficult point to achieve and it depends on serious science-oriented political leadership and committed international technological cooperation by engaging Arctic and non-Arctic nations in an aligned sustainable development process, because, in the end, whatever happens to the Arctic will affect the whole planet.

Indigenous communities in the resource-rich areas of the Arctic are increasingly exposed to a range of pre-existent risks, severe climate change impacts as well as the external pressures of development advocated by governments and its industry partners. With the discovery of vast energy and mineral resources in previously inaccessible areas of the north, the governments of Arctic littoral states are taking political measures to assert their territorial sovereignty over frozen land and the newly opened waterways. Modern infrastructure is an essential part of the planning of the remote settlements energy development in order to create favourable conditions of energy generation, environmental protection and social welfare.

At a local perspective, Arctic ecosystems are intrinsically diverse, vulnerable and dynamic. Most part of them are highly productive providing essential ecological services for other interdependent ecosystems and human communities. A number of Arctic marine and terrestrial ecosystems are also interdependent and have been stressed by locally- and globally-produced pollution from different geographical areas and sectorial sources since the start of the industrial revolution. Transboundary pollution produced by natural resources exploitation and transformation processes added to the effects of climate change have both global and local impacts on ecosystems and human communities.

A low carbon energy future not only depends on the way we generate and use energy, but also on how we access, transmit, and connect different systems. Smart grid technology involves and provides decentralization, flexibility and integration as essential characteristics towards a sustainable energy system transformation.

Decentralization and integration are core components of being 'smart' as a grid. Decentralization of electricity systems consists in energy users not being passive purchasers of electricity from utility companies. Consumers became 'prosumers' (Energy.gov, 2017) as they control electricity production, storage and transmission as well as they, proactively, manage their energy consumption through smart meters and energy efficient appliances. Consumers have the power of making daily educated decisions on how and when electricity is used depending on more or less abundant resources at certain periods of time to avoid peak demand periods.

This decentralization feature embeds the concept of 'efficiency and self-management' an extremely important innovative mind-set to promote and operate sustainable energy systems in a very sensitive environment like the Arctic. This mind-set is aligned to the 'triple bottom line' (Elkington, 2004) in the sense it stimulates a more balanced approach regarding social, environmental and economic components of sustainability. This makes the consumer dynamically integrated and a participant stakeholder in the energy supply-and-demand chain.

Integration means high interconnectivity of a wide and flexible range of renewable energy sources, including intermittent sources like photovoltaic solar, wind, with more predictable and controlled sources like geothermal, hydropower and biomass. A variety of energy customer services, energy storage systems and small-scale heat and power systems under an interconnected electricity platform contribute to higher interconnectivity (World Energy Council, 2012). This flexible model of interconnectivity also occurs through different interfaces linking residential, commercial and industrial energy management systems, via smart metering infrastructure made of automated communication platforms that also make the system more independent, self-managed and cost-effective. The energy applications provided by the integrated energy platforms can also work more efficiently not only by connecting different energy sources, but also linking government, businesses, consumers and industries thus making them part of an intelligent integrated power system where supply and demand can be flexibly customized, efficiently managed and better controlled.

Another fundamental aspect for building a new energy perspective is the application of concrete measures of energy efficiency. We believe that the reduction of fossil fuels can only be possible by associating renewable energy with energy efficiency improvement options like passive housing (PHIUS, 2018), passive building standards of design by maximizing the use of high performance materials and energy loss strategies. Efficiency measures can exponentially reduce energy demand in households and various industries as discussed in Chapter 1.

This intelligent energy system started being applied to Alaskan remote rural Arctic communities since 2008 when the Renewable Energy Fund was established with the intent of providing stable, reliable and affordable energy. From 2008 to 2015, approximately, US$257 million was made available for Renewable Energy Fund (REF) projects that also counted with local governments support and funding to reduce the energy costs associated to electricity generation and heating in Alaskan communities. In 2016, 70 operational projects displaced the equivalent to 31 million gallons of diesel fuel across the communities served by the new system. In terms of carbon emissions REF projects eliminated 857,875 metric tons of CO_2 from 2009 to 2016 boosting the renewable energy industry in Alaska (AEA, 2016).

Alaska is at the front in terms of the number of mini-grid projects and technology applied. This is the result of the combination of local technology developed by energy centres in important knowledge clusters plus the remarkable programmes and funding schemes of the Alaskan Energy Authority (AEA) that had an extraordinary social impact in more than 150 remote communities (AEA, 2016) providing energy access at affordable costs as well as significant amounts of carbon offset.

Joint assessments and social consultation on impacts were performed to help develop a new model of energy resources co-management for the region based on the interactions of experience-based knowledge from communities and scientific knowledge provided by teams of experts from local knowledge clusters. This co-management model provides mechanisms of managing benefits and critical impacts like low rates of energy access, equipment maintenance and environmental impacts on local biodiversity, pollution generation and land use.

These reasons justify an integrated and multidisciplinary approach to modern Arctic energy management. Especially considering the whole range of Arctic eco-systemic components, including human communities in Arctic towns and in remote rural areas, which would contribute to tackling the developmental complexity by identifying key areas of protection and their respective structures of functioning to provide important eco-systemic resilience and smart energy solutions.

Some examples of smart energy solutions currently applied in Alaska involve the application of the latest renewable energy technologies in small and large hydroelectric facilities, wind generation, geothermal pumps, air heat pumps, and woody biomass for electricity and heating (CCHRC, 2016; REAP, 2016). In addition, forms of renewable energy that seemed improbable in the past, like solar electricity generated from photovoltaic (PV) panels, are increasingly being analysed and deployed in Alaska due to specific meteorological factors like low temperatures that improve the efficiency of solar modules, the reflectivity of sunlight from the snow cover as well as the long periods of sunlight during summer season.

Wind power generation is another important resource with huge potential in Alaska due the strong winds mainly in the coastal regions. It has been widely disseminated since 2012 when the largest projects started in locations like Kodiak with six turbines and a total capacity of approximately 9 MW; Fire Island with 11 turbines and a total capacity of approximately 18 MW; and Eva Creek counting on 12 turbines with a total capacity of approximately 25 MW at a cost of $3,780 per installed kW with a total cost of $93,000,000 (REAP, 2016).

In general, the turbines installed in Alaska are of three specific sizes:

a   the small wind turbines (<10 kW) e.g. Skystream 3.7–2.4 kW;
b   intermediate (10–250 kW) e.g. Northwind 100–100 kW;
c   large (250 kW–6 MW) e.g. GE 1.5–1,500 kW.

The turbine sizes fit different areas where the AEA undertook wind study reports bringing on board Meteorological Data showing the wind power class essential to calculate the average net power output, the annual net energy output and, consequently, the average net capacity factor that defines the size of the wind power project to be implemented.

The Arctic micro-grid operations need to be understood in terms of possibilities and limitations. Under these lenses, the micro-grid systems in Alaska and Canada reflect an existing territorial energy system created by local and national partnerships trying to tackle unprecedented challenges in modernizing the off-grid rural energy infrastructure. In remote communities of Alaska and Canada rich in renewable energy sources, the intermittent nature of renewables makes their integration into the power grid a real challenge that requires proper stabilization technology enabling high penetration of renewable power generation, distribution control systems provided by smart electricity management and efficient hybrid power generation to maximize diesel savings of back-up diesel generators.

These leading initiatives on integrating renewable energy sources to the micro-grids in Alaska have mobilized efforts translated into the 'Pathway for Holistic Community Microgrid Development' by the Grid Modernization Laboratory Consortium (GMLC, 2014) consisting in a systematic set of actions to 'Assess, Conduct, Analyse, Consider, Design, Finance and Share' with the main aim to reduce fossil fuel energy use. The starting point is to assess the community capacity in order to conduct the infrastructure adaptation. The next step

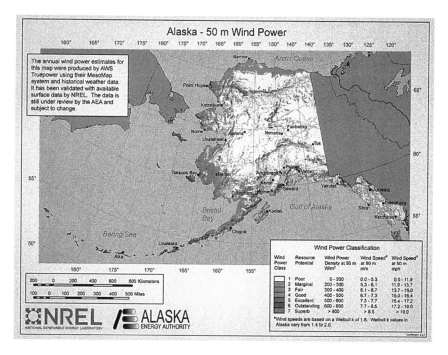

Figure 5.4 Alaska 50 m wind power.

Source: NREL and AEA.

consists in analysing the opportunities by applying the existing analytical tools in order to determine what new technologies to consider when designing the projects and systems to perform the goals. The analysis also contributes to quantifying the necessary funds and the financial model to be applied by sharing information to attract project developers as well as public and private capital to implement the projects.

The technological and innovative energy approaches in Alaska not only have shifted energy production and use from fossil fuels to a more sustainable direction, but also have provided a new horizon of durable energy access to Alaskan rural communities.

Towards sustainable energy systems through smart grids in the Arctic

Alaska presents a unique energy matrix and infrastructure as none of its grids are connected to the rest of the US and many areas are not connected by roads and only accessible by air or sea. Alaska is home to the largest oil field in North America, the Prudhoe Bay on the North Slope that presently generates around 300,000 barrels per day of the more than 1.6 million barrels per day in 1988, showing a clear decline in oil reserves and production. Despite the significance of Prudhoe Bay production being around 20 per cent of the whole US oil production for more than three decades, the cost of energy in most Alaskan communities has always been extremely high (AKRDC, 2017; BP Alaska, 2016).

According to the Alaska Arctic Council Host Committee (AACHC) (2016), the average cost of electricity in rural communities is about $0.55 per kWh reaching $1 per kWh in some Alaskan locations while heating fuel ranges from $3 to $10 per gallon. The urban areas of the southern Alaska are served by a more structured and connected infrastructure of links like the Railbelt urban power grid as well as gas and hydroelectric sources meaning that the reality in the urban centres is distinct from the rural remote communities.

Alaska proved to be rich in renewable resources with different characteristics varying from hydro projects in the southern areas where the climate is warmer and rainy, to plenty of wind resources in areas like Kotzebue, Kodiak, Chukchi Sea and Bearing Sea. In southeast Alaska, the main renewable source, showing relevant economic feasibility, is the technology of heat pumps that utilizes heat from the air, ground or ocean.

Emerging technologies are also in the government's sights. Geothermal, wave and tidal potential, and river hydrokinetics have also been part of new research initiatives to enhance their economic viability considering local characteristics and long distances of transmission and distribution systems.

Solar and wind energy projects are spread out across the state in a more consistent way due to the efforts of the AEA since 2008 when the potential areas started to be mapped demonstrating the economic viability and competitiveness of the projects.

Figure 5.5 represents the trend in wind-installed capacity in Alaska since 2003. After 2008 there was a significant increase in the number of projects

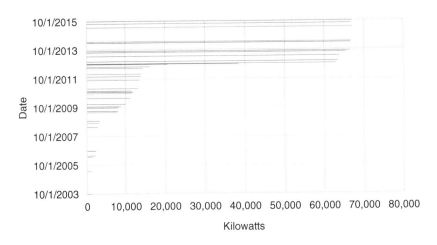

Figure 5.5 Installed wind capacity in Alaska.

Source: Anvivo.org.

providing more than 10,000 kW (Fay *et al.*, 2012). The most expressive projects were developed in Kotzebue, with an installed capacity of 66,644 kW, St George with the capacity of 66,739 kW, Buckland with 66,939 kW and Deering with 67,038 kW. There are other important wind farm projects under construction in Saint Mary's/Pitkas Point/Mountain Village, in Bethel, Chefornak and Kipnuk, all of them representing more than 68,000 kW (ACEP, 2012).

Rural Alaska has been part of a wide remote electrical grid modernization focused on minimizing the use of diesel generators. Currently, 70 out of 250 communities in Alaska are powered in part through renewable energy due to integration initiatives involving local micro-grid projects (REAP, 2016).

Micro-grids are self-contained local electrical networks that have one or more generation units feeding the local loads through a small network. The network operated under a low-medium voltage system (1,000 V or below) with small-scale combined renewable energy sources like solar PV, and wind turbines, that utilize heat, power and energy storage devices (Cheng, 2014). A micro-grid is a discrete energy system consisting of distributed energy sources (including demand management, storage, and generation) and loads capable of operating in parallel with, or independently from, the main power grid.

The Remote Community Renewable Energy Partnership and the Alaska Microgrid Partnership are important initiatives that have renewed the energy perspective of the Alaskan Arctic. Introduced by the National Renewable Energy Laboratory (NREL), as a national laboratory of the US Department of Energy (DOE) through its Office of Energy Efficiency and Renewable Energy, the laboratory operated by the Alliance for Sustainable Energy LLC was launched in November 2014. This was done under the US DOE's Grid Modernization Initiative, a strategic partnership between the US national laboratories and DOE headquarters, bringing together leading experts and resources to collaborate on national grid modernization goals (GMLC, 2014; NREL, 2011, 2013).

The GMLC had as a goal to coordinate US energy organizations that have been working on power systems for Alaskan communities and to develop accessible technological innovative systems in order to achieve at least 50 per cent reduction in the total fuel consumed in isolated communities. The current renewable energy deployment aims to reach more than 200 remote coastal and interior communities by installing wind-solar-diesel systems with the capacity of around 30 per cent of renewable energy system penetration in order to address the costs of electricity ($1.00/kWh) and diesel fuel (approximately $12.00 per gallon) varying by region.

The Consortium focused on six technical thematic areas viewed as essential to the adequate modernization of the rural Alaskan communities' energy infrastructure. The project involved the development of Technological Energy Devices (Wind Turbines, PV Panels and smart meters), the testing and measurements of electricity demand and supply, Systems Operations and Control, Design and Planning, Security and Resilience and Institutional Support to the areas where projects have been developed (Federal Energy Regulatory Commission Staff Report, 2006).

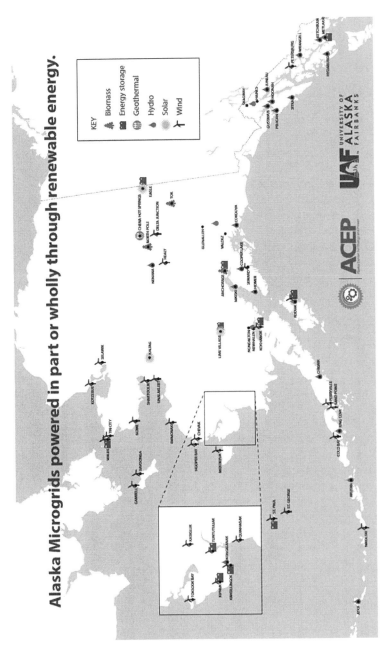

Figure 5.6 Alaska microgrid powered in part or wholly through renewable energy.

Source: ACEP.

The Remote Communities Renewable Energy micro-grids present challenges in relation to cost-efficiency as the renewable energy contribution of the systems accounts for approximately 30 per cent and the upfront investment continues to be significant even considering the decrease in prices of PV panels and wind turbines. Another important challenge is of a technical order as the renewable energy hybrid systems are highly complex and need to be adequately customized by engineers to the specificities of different locations. Additionally, the costs of communication and measuring devices and IT systems are relatively expensive and it is still difficult to provide accurate data related to systemic performance as no standard framework exists (Yu *et al.*, 2011). In fact, it is a work in progress that depends on funding and investment in technology development as well as in measurement and impact assessment.

Renewable energy micro-grid projects also face challenges related to economic and institutional aspects of financing, construction, operation and maintenance as there is a lack of trained workforce to install, repair and deal with the renewable energy systems technical complexities in remote Alaskan villages. There is also the need to develop customized business models for the sector, risk management strategies and industry standards due to the fact that project management costs continue to be high and require further institutional engagement with more advanced techno-economic analysis.

Remote Community Renewable Energy Partnership (RCRE) is an initiative funded by the US Department of the Interior aiming at developing opportunities to create more standardized, modular hybrid power systems and providing sustainable energy for rural Alaskan communities. The Alaska Microgrid Partnership is responsible for addressing the challenges by analysing case studies of pilot projects in Alaskan communities and by providing technical support developed by the branches of the US national Energy Labs. The Partnership also coordinates with the DOE initiatives together with the Office of Indian Energy Strategic Technical Assistance Response Team (START) and the Grid Modernization Alaska Microgrid Partnership (AMP) (GMLC, 2014).

The development of micro-grids for the Alaskan communities is far beyond being just a part of a modern trend; it represents the so-dreamed 'energy access' via technological innovation, local know-how and expertise being applied to the society. The energy flexibility and integration provide social and cultural interconnectivity and the benefit of creating smart communities defined by the Ministry of Economy, Trade and Industry (METI) (2013a, 2013b) as those that are aiming for efficient energy generation through utilization of innovative technology of renewable energy hybrid systems. These include cogeneration systems, heating from clean energy systems and energy storage facilities. These systems must manage both electricity and heating in a smart manner in order to achieve optimum efficiency of energy generation and use. Maturing smart communities depends on the integration of energy management systems with standards, data and core business systems.

The smart community represents a new type of societal organizational system in which its members are able to utilize a decentralized-type of energy, i.e.

renewable energy and cogeneration systems, and also access energy management systems taking advantage of advanced technologies, storage and metering systems. Under this framework, there is a significant contribution to manage regional energy demand, optimize energy use, minimize fossil fuels impacts and create a low carbon society.

The smart grids in the Arctic are an extremely relevant example of socio-technical energy systems. The concept of socio-technical energy system consists of how technologies and social responses evolve together and how their co-evolution affects planning and energy policies (Ali and Yekang, 2016). It also implies how the dynamics of new energy systems and associated social responses affect local systems of energy infrastructure, the built environments, and their residents in dealing with co-evolving technical and social dynamics.

The energy transition involves a social transition as technical and social dynamics co-evolve. This co-evolution is possible when technological and societal responses evolve together as the developments in smart technologies enables the creation and operation of new models of engagement between users and suppliers.

The relevance of this concept is the innovation in the mind-set oriented to the balance of economic, environmental and social pillars of sustainability in relation to energy systems. This reveals the importance of the human dimension and the community needs in designing local energy solutions for remote communities that primarily need energy access for heating and power. The contemporary perspective of energy transition informs social transition as the human dimension of technology and innovation.

Conclusion

A mixed institutional strategy of adaptation and mitigation has triggered an operative strategy to minimize the impacts of energy production and use in the Arctic. The deep understanding of Arctic climatic dynamics is crucial to implement this strategic process, as the multicultural understanding of each stakeholder's role in the energy development project allows them to become active parties and beneficiaries of development and prosperity. Indigenous communities must be an integral part of the energy transitions and sustainable development project based on shared values, traditional knowledge and co-management.

The grid of the future will be smart. It will integrate conventional and renewable sources with adequate and efficient energy storage systems and it will deliver reliable, sustainable and affordable electricity to consumers in city centres and remote areas. This is definitely the vision.

Energy transition is also a transition of socio-technical systems. Smart grids have a major role to play in a low carbon future in relation to energy saving, integrating a broad range of generation and storage options including renewable energy that is key to both demand-and-supply-side management of energy systems.

Increasing the share of renewable energy in total power generation will not only reduce carbon emissions and mitigate climate change but it will also reduce energy generation costs as well as energy demand with a significant impact on the socio-economic development worldwide.

References

AACHC (Alaska Arctic Council Host Committee) (2016) 'Environmentally Responsible Resource Use and Development in the US Arctic', Alaska Arctic Council Host Committee, October 2016, p. 14.

ACEP (Alaska Center for Energy and Power) (2012) 'An Investigation of Solar Thermal Technology in Arctic Environments: A Project by Kotzebue Electric Association', Denali Commission—Emerging Energy Technology Grant Report (2012).

AEA (Alaska Energy Authority) (2016) 'Annual Report', www.akenergyauthority.org/Portals/0/Publications/AEA2016AnnualReport.pdf?ver=2017-03-01-180130-100, p. 16, accessed 1 November 2017.

AKRDC (Alaska Resource Development Council) (2017) 'Alaska's Oil and Gas Industry', www.akrdc.org/oil-and-gas, accessed 10 December 2017.

Ali, A. and Yekang, K. (2016) 'Socio-technical evolution of Decentralized Energy Systems: A critical review and implications for urban planning and policy', *Renewable and Sustainable Energy Reviews*, Vol. 57, Issue C, pp. 1025–1037.

Arruda, G. M. and Krutkowski, S. (2017) 'Social impacts of climate change and resource development in the Arctic: Implications for Arctic governance', *Journal of Enterprising Communities: People and Places in the Global Economy*, Vol. 11, Issue 2, pp. 277–288.

BP Alaska (2016) 'Environmental Site Report', www.bp.com/content/dam/bp/en/corporate/pdf/sustainability-report/verified-site-reports/bp-alaska-site-report-2016.pdf, accessed 10 December 2017.

CCHRC (Cold Climate Housing Research Center) (2016) 'Annual Report', www.cchrc.org/sites/default/files/docs/2016AnnualReport.pdf, accessed 27 November 2017.

Cheng, J. W. M. (2014) 'A Holistic View on Developing Smart Grids for a Low-Carbon Future' in Mah, D., Hills, P., Li, V. O. K. and Balme, R., *Smart Grid Applications and Developments*, 1st edition, Springer, London. p. 29.

Cunsolo Willox, A., Harper, S. L., Ford, J. D., Landman, K., Houle, K., Edge, V. L. and the Rigolet Inuit Community Government. (2012). 'From This Place and of This Place: Climate Change, Sense of Place, and Health in Nunatsiavut, Canada', *Social Science and Medicine*, Vol. 75, Issue 3, pp. 127–147.

Durbha, S. S., Kuldeep, R. K. and Bhangale, U (2017) 'Semantics and High-Performance Computing Driven Approaches for Earth Observation (EO) Data-State of the Art', Indian Institute of Technology, Mumbai, India.

Electricity Advisory Committee (2008) 'Smart Grid: Enabler of the New Energy Economy', The Electricity Advisory Committee, US Department of Energy, Washington, DC, http://energy.gov/sites/prod/files/oeprod/Document-sandMedia/final-smart-grid-report.pdf, accessed 10 September 2017.

Elkington, J. (2004) 'Enter the Triple Bottom Line'. In Henriques, A. and Richardson, J., *The Triple Bottom Line: Does It All Add Up? Assessing the Sustainability of Business and CSR*, Earthscan, London, p. 23, 24.

Energy.gov (2017) 'Consumer vs Prosumer: What's the Difference?', US Department of Energy, Office of Energy Efficiency & Renewable Energy, https://energy.gov/eere/articles/consumer-vs-prosumer-whats-difference, accessed 30 June 2017.

ESA (European Space Agency) (2017) 'Earth from Space', www.esa.int/Our_Activities/Observing_the_Earth/Copernicus/Sentinel-2, accessed 20 November 2017.

ETPSG (European Technology Platform SmartGrids) (2006) 'Vision and Strategy for Europe's Electricity Networks of the Future'. Directorate-General for Research Sustainable Energy Systems. European Commission.

Fay, G., Villalobos Meléndez, A., Saylor, B. and Gerd, S. (2012) 'Alaska Energy Statistics 1960–2008', prepared for the Alaska Energy Authority. University of Alaska Anchorage. Anchorage, AK: Institute of Social and Economic Research, www.iser.uaa.alaska.edu/Publications/AlaskaEnergyStatisticsCY2008Report.pdf, accessed 20 November 2017.

Federal Energy Regulatory Commission staff report (2006) 'Assessment of Demand Response and Advanced Metering (Docket AD06-2-000 pdf)', United States Department of Energy, p. 20, accessed 27 November 2017.

GMLC (Grid Modernization Laboratory Consortium) (2014) 'Alaska Microgrid Partnership Project 1.3.2.1 Fact Sheet', US Department of Energy's Grid Modernization Initiative, https://gridmod.labworks.org/sites/default/files/resources/1.3.21_Alaska%20Microgrid%20Partnership_Fact%20Sheet_rev2.pdf, accessed 28 November 2017.

Ministry of Economy, Trade and Industry (METI) (2013a) 'ANRE's Initiatives for Establishing Smart Communities', www.meti.go.jp/english/policy/energy_environment/smart_community/pdf/201402smartcomunity.pdf, accessed 21 December 2017.

Ministry of Economy, Trade and Industry (METI) (2013b) 'Smart Community', www.meti.go.jp/english/policy/energy_environment/smart_community/index.html, accessed 21 December 2017.

NETL (National Energy Technology Laboratory) (August 2007) 'NETL Modern Grid Initiative — Powering Our 21st-Century Economy' (PDF). United States Department of Energy Office of Electricity Delivery and Energy Reliability, p. 17, accessed 10 October 2017.

NETL (National Energy Technology Laboratory) (2017) 'Modern Grid Initiative', US Department of Energy, www.netl.doe.gov/moderngrid/opportunity/vision_technologies.htmlArchived, accessed 1 September 2017.

National Snow and Data Center (2017a) 'Arctic Sea Ice News', http://nsidc.org/arctic-seaicenews/about-the-data/, accessed 27 December 2017.

National Snow and Data Center (2017b) 'Record Low Extent in the Chukchi Sea', http://cires1.colorado.edu/websites/nsidc/publications/index.php, accessed 27 December 2017.

NREL (National Renewable Energy Laboratory) (2011) 'CREST Cost of Energy Models', *Renewable Energy Project Finance*, National Renewable Energy Laboratory, https://financere.nrel.gov/finance/content/crest-cost-energy-models, accessed 20 December 2016.

NREL (National Renewable Energy Laboratory) (2013) 'Renewable Energy in Alaska', WH Pacific, National Renewable Energy Laboratory, www.nrel.gov/docs/fy13osti/47176.pdf, accessed 5 June 2017.

PHIUS (Passive House Institute US) (2018) 'Passive Building', www.passivehouse.us, accessed 20 December 2017.

Rao, U. R. (1991) 'Remote Sensing for Sustainable Development', Vikram Sarabhai memorial lecture delivered at the annual meeting of the Indian society of remote sensing at Madras on 11 December 1991. Indian Space Research Organisation, Bangalore.

REAP (Renewable Energy Atlas Project) (2016) 'Renewable Energy Atlas of Alaska. A Guide to Alaska's Clean, Local and Inexhaustible Energy Resources', Alaska Energy Authority, April 2016.

Roy P. S. and Behera M. D. (2000) 'Perspectives of Biodiversity Characterization from Space', *Employ News* (Gov India), Vol. 25, Issue 16, pp. 1–2.

Roy, P. S., Behera, M. D. and Srivastav, S. K. (2017) 'Satellite Remote Sensing: Sensors, Applications and Techniques', *Proceeding of the National Academy of Science India, Section A Physical Sciences*, Vol. 87, Issue 4, pp. 465–472.

USGS (US Geological Survey) (2017) 'What are Climate Data Records and Terrestrial Essential Climate Variables?', US Department of the Interior, https://remotesensing.usgs.gov/ecv/index.php, accessed 7 July 2017.

World Energy Council (2012) 'Smart Grids: Best Practice Fundamentals for a Modern Energy System', World Energy Council, www.worldenergy.org/documents/20121006_smart_grids_best_practice_fundamentals_for_a_ modern_energy_system.pdf, accessed 11 July 2017.

Yu, F. R., Zhang, P. W. X. and Choudhury, P. (2011) 'Communication Systems for Grid Integration of Renewable Energy Resources', *IEEE Network*, Vol. 25, Issue 5, pp. 22–29.

6 Arctic resource development

A sustainable prosperity project of co-management

Gisele M. Arruda

Introduction

It seems exciting to envisage the contemporary Arctic development process and its human dynamics. However, it is important to realize what is below the top of the 'iceberg'. It may hide a natural resources rush that can have dramatic consequences for the development of the region.

The end of the Cold War has triggered a new process for the Arctic region as it became less militarized and a new space was opened for entrepreneurship and economic development stimulated by political support on industrialization based on the abundance and the attractive prices of local commodities. These factors increased interest in industrial and maritime activities in the region representing an incipient and unregulated reality for the local context. Industrial and maritime activities are synonymous with energy demand and risk, apart from the own risks imposed by Climate Change. Local environmental and social risks have been materialized into Persistent Organic Pollutants (POPs), heavy metals, hydrocarbon pollution, methane emissions, wastes pollution and radioactivity that are common factors affecting ecosystems and human health.

This chapter aims to raise the debate about the implications of resources development and industrialization process in the Arctic in order to explore some pathways for a future prosperity project which will certainly not be performed as the business-as-usual, under the current governance regime.

Arctic-specific natural ecosystems, the presence of Indigenous communities and the commercial interest in the region present a layer of risks that is asseverated by specific risks posed by Climate Change. This complex range of pre-existent and new risks will certainly require an innovative model of development and governance based on the highest level of responsible exploitation, diplomacy, regulation and policy-making. Probably, this political formula has not been shaped yet, but it may possibly become a collective construction based on co-management and multilateral cooperation of Arctic and non-Arctic nations for a common well-being.

This is an ongoing process. A moment in history when leaders have the opportunity to choose the way to take, the formulas to be applied, the innovations to promote, not only in technology but in the developmental mindset. The chapter employs a review of past and recent literature, face-to-face interviews with representatives of local communities and entrepreneurs. It is also the result of brainstorming with peers over the last three years and a means to discuss present issues and future management strategies for designing an innovative model of Arctic development and governance based on co-management and multicultural dialogue in an attempt to achieve convergence.

Climate Change and modernization

The Arctic plays a key role in the global climate. As the ice retreats the current challenges concerning the environment, maritime safety, tourism, and oil and gas activity will intensify even more the effects of Climate Change on ecosystems and communities. Low carbon transitions are the product of wider changes in the geographic organization of economic activity, however, they are structured by the actions of firms and governments acting at multiple scales (Smil, 2005, 2010).

There are also ethical dilemmas involved in this analysis. Modernity has in itself 'construction' and 'destruction' what makes 'modernity' an intrinsic paradox of creation and deconstruction, presenting us a serious dilemma and existential crisis. Modernity can, on one side, bring on board powerful technologies to promote well-being and at the same time it can provoke a profound crisis of reality and truth. The results and fruits of modernity can be distinct for different societies, depending on their relationship with the models of development created by global and local players.

Indigenous communities in the resource-rich areas of the Arctic are increasingly exposed to a range of pre-existent risks, severe Climate Change impacts as well as the external pressures of development advocated by governments and its industry partners. With the discovery of vast energy and mineral resources in previously inaccessible areas of the North, the governments of Arctic coastal states are taking new measures to assert their territorial sovereignty over frozen land and the newly opened waterways. Modern infrastructure is part of the planning of the remote settlements in order to facilitate the logistics of the exploration.

Climate Change and modernization have thus become two intrinsically linked forces that severely alter the context in which the Indigenous populations of the region sustain a livelihood (van Voorst, 2009). The latest ice modelling results produced by polar-orbiting satellites revealed factual evidence of significant reduction in the Arctic Sea ice since 2010 and massive ice losses in the central part of Greenland tending to asseverate until 2020. Land ice losses represent open areas for 'extractivism' and sea ice loss represents new prospects for offshore oil production, mining and new marine routes available. Modernization is perhaps the main driver of all changes synthesizing causes and effects of physical and societal transformations.

In an era of globalization, the conventional paradigm of economic policy is in need of radical rethinking. Such a paradigmatic shift, however, will necessarily have to be accompanied by practical efforts to re-embed the global economic system in qualitatively new social relations and forms of political regulation, on both local and global levels (Altvater, 2002).

We need to be particularly concerned about unilateral utopian schemes for a society to resolve the problems of modernity, mainly when these schemes reproduce the same inefficient discourse from the past. In the Indigenous cultural context it will not be an easy task to conciliate a subsistence-hunting local economy with a Westernized market economy stratification. Modernization provides the benefits of a petroleum-based model of development, with the associated range of by-products and facilities the Western societies consume. However Indigenous populations have their native operative model based on local perception and interpretation of nature, land, livelihood, skills, subsistence, spirituality, work and well-being.

Indigenous people have evolved ways of living that are well suited to vulnerable environments. Western culture has a different perspective in relation to the carrying capacity of particular lands. There is a belief that any land can be put to use and not wasted for subsistence economic activities. Land use regimes seem to be a fundamental source of conflicts that will need to be addressed by managers, governments and Indigenous leaderships.

Probably, the first step to build up a sustainable development vision for the Arctic in order to progress towards a sustainable project of co-management may start with the multicultural understanding of the concept of development, by Indigenous Arctic communities and Western societies.

The Western concept of sustainable development has gathered strength from a variety of international declarations, conventions and academic production. The term 'sustainability' that is the function of Sustainable Development, entered into common usage relatively recently following the publication of the report 'Our Common Future' by the United Nations, Brundland Commission in 1987. The Commission defined sustainability and, in particular, sustainable development as 'Development that meets the needs of the present generation without compromising the ability of future generations to meet their needs' (United Nations, 1987). The Triple Bottom Line framework introduced in the mid-1990s incorporates three dimensions, commonly known as economic, social, environmental or the so-called three Ps: people, planet and profits.

The Indigenous perspective of development has to do with the meaning of place where they have conditions to be 'Indigenous' where they are able to earn a livelihood in their own land, to practice traditional hunting and fishing, where to cultivate their social and family relations, and practice their rituals and healing.

Socio-environmental impacts at local and global levels

Environmental conditions, geological potential and accessibility, population rates, economic development and political leadership are important variants that determine different risk levels in different territories in the Arctic. The balance of risk and opportunity for different models of development is a very difficult point to achieve and it depends on serious political leadership and committed international cooperation by engaging Artic and non-Arctic nations in the development process, because, in the end, whatever happens to the Arctic will affect the whole globe.

Transboundary pollution produced by natural resources exploitation and the effects of Climate Change have both global and local impacts on ecosystems and human communities. Locally speaking, Arctic ecosystems are intrinsically diverse, vulnerable and dynamic. Most of them are highly productive, providing essential ecological services for other interdependent ecosystems and human communities. A number of marine and terrestrial ecosystems are also interdependent and have been stressed by locally- and globally-produced pollution from different geographical and sectorial sources since the start of the industrial revolution.

The Arctic seems to be vulnerable to local and foreign pollution due to the marine and air currents, low temperatures and geographical characteristics. According to Arctic Monitoring and Assessment Programme (AMAP) Arctic Pollution Issues 2015 (AMAP, 2015), the most serious environmental stressors in the Arctic are POPs, brominated flame retardants (BFRs), polyfluoroalkyl substances (PFASs), chemicals used in pesticides, heavy metals and radioactivity, despite the limited human development levels in Arctic areas. POPs are long-lasting chemicals that pose health risks to ecosystems and human communities. They are transported long distances and deposited far from their sources of release as configuring a classic case of transboundary pollution. They tend to accumulate in the fatty tissues, milk and blood of living organisms compromising the local food chain and affecting other dependant ecosystems and human health in the Arctic local communities.

Heavy metals like mercury and methylmercury, originated from mining activities from outside the Arctic Circle, are also a threat to Arctic ecosystems and local communities as a result of the cumulative effect they have in the food chain. Radioactivity is an additional great concern as a result of nuclear tests, transfer pathways and inadequate waste management since the 1950s.

Arctic pollution levels have been monitored by AMAP since the 1990s, demonstrating that levels of POPs declined over the past 30 years, while BFRs and PFASs have increased and are a matter of concern as mercury levels and anthropogenic levels of radioactivity remain high according to the latest AMAP assessments of 2017 (AMAP, 2017). Pollution monitoring activities in the Arctic have been based so far on a list of chemicals in extensive use, for more than a century, in the European Union (EU) and the US, but the great concern is on potential contamination originated from unmonitored sources which impacts remain unknown.

In Greenland, the Cooperative Sheep Farmers Associations have written an open letter to the Greenlandic government, requesting that radioactive wastes from the exploration of rare earths and uranium is not thrown into the lake at Narsaq during mining operations that are planned to start in two years' time in the region. The majority of sheep farmers expressed their concern about the management of radioactive wastes in the settlement, but they do not believe that only being consulted will prevent the occurrence of pollution. The same situation is seen in Kuannersuit in Canada where SIK Mining believes that stopping uranium extraction in that location would prevent the development of Canada. The difference between the first and the second case is that, in Canada, there is a legal framework about uranium production but in Greenland there is no legislation in place for extraction and export of uranium that drastically enhances the operational risks in comparison with the benefits awarded by this kind of activity. In both cases, the argument for moving the exploration forward is the economic appeal of jobs creation, but how long will economic development be the argument for social and environmental degradation? How long will the social and environmental components of the Triple Bottom Line continue to be ignored?

Traditional lands have been subject to a range of development activities including hydroelectric and irrigation dams, logging, pulp and paper mills, mining and tourism. Hydroelectric projects were reported to have affected huge areas in the Indigenous territories that were once traditional hunting and fishing areas. Mercury contamination from mining activity has made commercial or subsistence fishing substantially dangerous in the North where a number of other case studies report mercury poisoning in the local population. The quality of water is under serious analysis due to infrastructural projects.

The traditional way of life seems to be permanently disrupted as social relations were altered and family systems were broken down due to the rapid Arctic transformation. The habits have also changed, showing high rates of alcoholism, tobacco addiction and suicides. Other violent deaths became common occurrences lately. Entire communities have been forced to relocate to undesirable areas demonstrating the severe impacts on the Indigenous traditional economy (Koivurova and Heinämäki, 2006). In general, we can affirm that conditions for Indigenous people worsened as they are unable to earn a livelihood from their traditional lands as a result of the competition for natural resources, energy production and because of the mismatch of their skills and abilities to gain employment from the new projects.

The 'Licence To Operate' (LTO) framework for the Arctic offshore exploration is not yet established in many different Arctic environments. Over the past decades, the most part of oil and gas operations were developed onshore. Traditional companies like British Petroleum (BP), for instance, started their exploration in the Arctic in 1959, in Alaska, moving subsequently to the North coastal region. Since 1977, BP has operated Prudhoe Bay field, producing more than 12 bnbblo (billion barrels of oil) and has known indeclinable capacity for at least another 35 years of production at the North Slope, accounting for inventories estimated at more than double that of Prudhoe Bay. BP currently operates in 15 North Slope oil fields with the opportunity to operate in six more new fields in the following five years. In terms of offshore exploration, in the 1980s BP made oil and gas discoveries in the Beaufort Sea and the Arctic Islands of Canada, with fields in operation since the 1980s until the present day (Daly, 2012).

In relation to the licensing process, the main challenge seems to be the national regulatory systems of the applicants that present a range of contradictions, ambiguities and discrepancies creating insecurity. The case of Russia offshore exploration is conveyed as an illustration of the controversies of its licensing process and legislation, due to the fact that Russia is currently a polarizing force because of Russia's Arctic Strategy which describes the Arctic zone as a 'national strategic resource base' representing the broader framework of Russian national policy (UArctic, 2014). In other words, the current Russian state policy is to expand the resource base and the military capacity of the Arctic zone, representing the main Russian strategy for national economic growth in the following 50 years.

A joint assessment and management on impacts issues must be performed based on mutual consent, information exchange, responsible coordination and cooperation. These components could inform a new model of resources co-management for the region founded on experience-based knowledge and scientific knowledge. This co-management model is expected to provide mechanisms of managing benefits and critical impacts towards communities such as pollution and land use.

As to climatic impacts, they refer to more than just temperature and precipitation. It includes extreme events, as well as aspects of the system such as snow, ice and circulation patterns in the atmosphere and oceans. In the Arctic, sea ice is one of the most important climatic variables. It is key indicator and agent of Climate Change, affecting surface reflectivity, cloudiness, humidity, exchanges of heat and moisture and the ocean surface, and ocean currents (AMAP, 2004).

Climate Change is rapidly transforming the Arctic environment at a dramatic speed, affecting the capacity of local communities to completely understand and adapt to the physical, social and environmental changes in place. New knowledge alone, without some change of feeling and purpose, will not suffice to make international cooperation the normal method of resolving conflicts in the Arctic region. The 'superstructures of law and policy' will need, at this point, to work beyond the foundations of authority to operate the quantification of environmental and human risks including components of justice and human rights (United Nations, 2015). This process will also need to count on Indigenous communities' participation to provide a mutual understanding of the problem and to make them part of the solution and, mainly, part of the regional development process.

The current institutions for monitoring and reporting of emissions created under the auspices of the United Nations Framework Convention on Climate Change (UNFCCC) and Kyoto Protocol laid important foundations to the current political and legal frameworks structural negotiations. The elements continue to be cooperation and the continuous adaptation of the own institutional structures to the challenges of mitigation and adaptation. Both mitigation and adaptation became part of the risk management requiring international frameworks to set mechanisms of accountability for local and global actions.

The institutional adaptation at this transitional time would require:

a   a clear understanding of ecological limits;
b   a more ecologically protective and cultural-oriented policy;
c   measuring the effects of energy policy according to social and environmental parameters;
d   implementing energy policy according to agreed norms, social values and regulation.

Risk assessment and environmental management

Development factors represent an important element for management. Human health and food supplies are commonly affected by environmental components and consequently there is a significant interest in assessing risks caused by change of climate, energy and mineral projects, pollution and environmental changes.

Risk assessment and management are essential mechanisms for the Arctic development process because it is fundamental to identify the limits and boundaries of socio-environmental impact and resilience for the region.

In recent years, new risk assessment reports have highlighted the impact of Climate Change on Arctic marine and terrestrial environments. A comprehensive report on ocean acidification in the region, released by the Arctic Council, confirms that among the world's oceans, the Arctic Ocean is one of the most sensitive to ocean acidification (AMAP, 2013), and that Arctic marine ecosystems are highly likely to undergo significant changes as a result. Another Arctic Council report, the 'Arctic Biodiversity Assessment', confirms that Climate Change is the most important stressor for Arctic biodiversity and will exacerbate all other threats. Increased human activities, such as oil exploration and shipping, will place additional stress on the region's biodiversity (UNEP, 2014; UNEP Yearbook, 2014).

The Arctic environment is not uniform because there are areas of immediately viable resources recovery, areas of potential resources extraction and zones of harsh environments where resources are unlikely to be extracted in the short and medium term. This is the origin of the jargon 'many Arctics'. The great contemporary challenge is to manage these different Arctics in terms of impacts, policies and regulation.

According to the latest reports from the oil and gas industry, no oil company to date is completely confident in its expertise and technology for the particular conditions of the Arctic. Based on this, we advocate the position that no Arctic offshore drilling should be approved until a number of outstanding issues are clearly addressed such as the standards for health and safety, oil spill prevention, and accurate response to incidents and accidents. A comprehensive environmental impact report for the targeted areas is required in order to ensure the protection of the Arctic marine ecosystem and communities. It is important to achieve a balance between energy development and socio-environmental protection through specific zoning and adequate regulation but these initiatives need to be intrinsically aligned to a Climate Change legal framework that seems not to be satisfactorily considered in the recent licensing process (Arruda, 2014).

Environmental management is a politicized process (Wilson and Bryant, 1997) mainly in times of globalization and interdependency. It requires a good level of international co-ordination and control. Transboundary and global management have grown in importance and impact over the last decade but are still weak in regards to enforcement and overall co-ordination of measures to protect society.

In any environmental management situation there are several different views and different possible responses but the environmental manager has to try to avoid conflicts between stakeholders and minimize damage to the environment. The environmental manager deals with policy, planning, legislation, control, management and implementation (Cooper, 1995). He should also be able to build up trust, influence opinions, establish supportive institutions, inform and create network of relations, build up bridges, and, in the end, strengthen the development process. Environmental managers seem to have an unprecedented role in the current context of challenges that comprehends dealing with the pre-existent risks and tackling Climate Change through the dissemination of information and providing support to the adaptation process globally and locally. Since the 1970s, environmental management characteristics have evolved to far beyond the solely formal roles of assessing and managing risks towards a multi-disciplinary process of dealing with human-environment interaction. Until 2030, environmental management will probably continue to experience a major adaptation to an impermanent reality.

An integrated and multidisciplinary approach to management considering the whole range of components of ecosystems including human communities would consist of tackling the developmental complexity by identifying key areas of protection and their respective structures of functioning to provide important ecosystems and social resilience.

The Arctic operations need to be understood in terms of possibilities and limitations but mainly in terms of socio-environmental pillars of sustainability.

Social Impact Assessments (SIAs)

Social Impact Assessments (SIA) can be defined as the process of assessing or estimating, in advance, the social consequences that are likely to follow from specific policy actions or project development, particularly in the context of appropriate national, state or provincial environmental policy legislation. Social impacts include all social and cultural consequences to human populations of any public or private actions that alter the ways in which people live, work, relate to one another, organize to meet their needs and generally cope as members of society. Cultural impacts involve changes to the norms, values and beliefs of individuals that guide and rationalize their cognition of themselves and their society. While SIA is normally undertaken within the relevant national environmental policy framework, it is not restricted to this, and SIA as a process and methodology has the potential to contribute greatly to the planning process (Burdge and Vanclay, 1996).

In general, SIA, according to the Interorganizational Committee (1994), is revealed to be an important tool to assist in the process of understanding and evaluating social changes and possible alternatives to tackle them. The pertinence of discussing SIA within the context of Arctic development and energy transition is central due to the level of changes associated to economic exploration of natural resources in the region as well as the level of decision-making desired in this process.

SIA process provides direction in (1) understanding, managing and controlling change; (2) predicting probable impacts from 'change strategies' or development projects that are to be implemented; (3) identifying, developing and implementing mitigation strategies in order to minimize potential social impacts (those ones that would occur if no mitigation strategies were to be implemented); (4) developing and implementing monitoring programmes to identify unanticipated social impacts that may develop as a result of the social change; (5) developing and implementing mitigation mechanisms to deal with unexpected impacts as they develop; and finally (6) evaluating social impacts caused by earlier developments, projects, technological change, specific technology and government policy (Burdge and Vanclay, 1996).

SIAs serve as a means of determining how and to what extent specialized social groups will become better or worse off as a result of certain externally generated actions. Assessments have been largely about Indigenous people, not by them (Cochran *et al.*, 2013).

It reduces the stress caused by uncertainty and maximizes community engagement in the process of development. This is why it becomes crucial to enrich SIAs with detail and context that focus on the Indigenous perspective, in which economy and culture are more closely intertwined.

The Corporate Social Responsibility (CSR) agenda has been pushed by international issues with the advent of globalization and the corporations' socio-environmental responsibilities. CSR for the Arctic represents the opportunity of a re-evaluation of the role of corporations in society, emphasizing the contribution business makes to society (Handy, 2002; Teach, 2005).

Corporate governance is concerned with holding the balance between economic and social goals and between individual and communal goals. The corporate governance framework is there to encourage the efficient use of resources and equally to require accountability for the stewardship of those resources. The aim is to align as nearly as possible the interests of individuals, corporations and society (Iskander and Chamlou, 2000).

This is why it becomes crucial to enrich SIAs with detail and context that focus on the Indigenous perspective, in which economy and culture are more closely intertwined. The benefits of the Arctic emerging economy may be seen in the creation of economic development, but it must be part of a sustainable prosperity project of co-management with triple gain to economy, environment and communities.

Recommendations for a sustainable prosperity project of co-management

A mixed strategy of adaptation and mitigation could be the beginning of an operative strategy for the Arctic, with the multicultural understanding of each stakeholder's role in the development project by making them active parties and beneficiaries of development and prosperity being crucial to this process. Indigenous communities must be an integral part of a development project based on shared values and traditional knowledge.

Government and political decision-making must operate closer to people's lives, and decentralization has been seen as necessary for a more democratic perspective of resource development, environmental management and prosperity creation.

The negotiating process should be designed to support proactive action from stakeholders by setting pre-commitments and reciprocal offers of cooperation in terms of emissions trading, technology transfer, transboundary pollution abatement and finance for adaptation. It is essential that Climate Change is fully integrated into development policies, providing conditions for collective governance.

World-class research, monitoring, information sharing on Climate Change impacts, risk assessment and adaptive environmental management according to multicultural approaches can help to reduce socio-environmental risks in sensitive areas. Adaptation strategies focused only on vulnerability, risk assessment and consultation procedures will not be enough. It is essential to rethink adaptation as an ongoing process, focusing on resilience and measurement of operating limits through effective and negotiated socio-environmental local and international coordinated strategies.

The current politics of Arctic development is controversial and vague considering the Arctic's iconic status and sensitive environment. Arctic development is often politically contentious, with opposing interests and perspectives between local, national and international levels. Political support for development needs to evolve from uncertainty for businesses seeking to invest in an Arctic development project to politics that are more informative than just a retouch of the superficial catastrophic discourse.

The Arctic transition requires the highest level of responsible development and governance standards ever seen on a global platform. The political arena can serve as a means of pacification, conciliation of interests and durable prosperity. In this sense, multicultural dialogue should be stimulated in pro of convergence of multidimensional interests from the Arctic and non-Arctic stakeholders.

The benefits of the Arctic emerging economy may be seen in the creation of economic development, but it must be part of a sustainable prosperity project of co-management with triple gain to economy, environment and communities. The development of a sustainable vision that incorporates the multicultural values of Indigenous polar communities would require, in practice, a democratic, collaborative format of governance inspired in authentic co-management.

References

Altvater, E. (2002) *Globalisierung der Unsicherheit. Arbeit im Schatten, Schmutziges Geld und informelle Politik.* Münster, Westfälisches Dampfboot.

AMAP (2004), 'Impacts of a Warming Arctic', Arctic Climate Impact Assessment (ACIA), Cambridge.

AMAP (2007) 'Arctic Oil and Gas 2007. Arctic Monitoring and Assessment Programme', AMAP, Oslo, Norway.

AMAP (2013) 'Assessment 2013: Arctic Ocean Acidification', AMAP, Oslo, Norway.

AMAP (Arctic Monitoring and Assessment Programme) (2015) 'Summary for Policy-makers: Arctic Pollution Issues', AMAP, Oslo, Norway.

AMAP (Arctic Monitoring and Assessment Programme) (2017) 'AMAP Assessment 2016: Chemicals of Emerging Arctic Concern', AMAP, Oslo, Norway.

Arruda, G. M. (2014) 'Global governance, health systems and oil and gas exploration', *International Journal of Law and Management*, vol. 56, no. 6, pp. 495–508.

Burdge, R. and Vanclay, F. (1996) 'Social impact assessment: a contribution to the state-of-the art series', *Impact Assessment*, vol. 14, no. 1, pp. 59–86.

Cochran, P., Huntington, O. H., Pungowiyi, C., Tom, S., Chapin III, F. S., Huntington, H. P., Maynard, N. G., and Trainor, S. F. (2013) 'Indigenous frameworks for observing and responding to climate change in Alaska', *Climatic Change*, vol. 120, no. 4, pp. 557–567.

Cooper, P. (1995) 'Toward a Hybrid state: the case of environmental management in a deregulated re-engineered state', *International Journal of Administrative Sciences*, vol. 61, no. 2, pp. 185–200.

Daly, M. (2012) Arctic Energy Agenda Roundtable, The Royal Norwegian Ministry of Petroleum and Energy, Trondheim, Norway, www.bp.com/en/global/corporate/press/speeches/the-arctic-a-future-licence-to-operate.html (accessed 25 June 2012).

Handy, C. (2002) 'What's business for?', *Harvard Business Review*, December, pp. 49–55.

Interorganizational Committee on Guidelines and Principles (1994) Guidelines and Principles for Social Impact Assessment, *Impact Assessment*, vol. 12, no. 2, pp. 107–152.

Iskander, M. R. and Chamlou, N. (2000) 'Corporate Governance: A Framework Implementation', The World Bank Group, Washington DC, US.

Koivurova, T. and Heinämäki, L. (2006) 'The participation of indigenous peoples in international norm-making in the Arctic', *Polar Record*, vol. 42, no. 2, pp. 101–109.

Smil, V. (2005) 'Limits to Growth revisited: a review essay', *Population and Development Review*, vol. 31, pp. 157–164.

Smil, V. (2010) 'Energy Myths and Realities: Bringing Science to the Energy Policy Debate', *American Enterprise Institute for Public Policy Research*, Washington DC, pp. 14–213.

Teach, E. (2005) 'Two Views of Virtue', *CFO*, December, pp. 31–34.

United Nations (2015) 'Achieving Sustainable Development and Promoting Development Cooperation', www.un.org/en/ecosoc/docs/pdfs/fina_08-45773.pdf, accessed 10 November 2016.

UArctic (2014) 'UArctic Strategic Plan 2020', www.uarctic.org/dm_documents/UArctic_Strategic_Plan_2020_FINAL_031213_MVPCx.pdf, accessed 1 November 2016.

UNEP (2013) 'The View from the Top: Searching for Responses to a Rapidly Changing Arctic'. In: *UNEP Yearbook 2013: Emerging Issues of our Environment.* UNEP Division of Early Warning and Assessment, Nairobi, Kenya.

UNEP (2014) 'Environmental Management in Oil and Gas Exploration and Production', www.ogp.org.uk/pubs/254.pdf, accessed 1 November 2016.

United Nations (1987) 'Report of the World Commission on Environment and Development', General Assembly Resolution 42/187, 11 December 1987.

van Voorst, R. S. (2009) 'I work all the time – he just waits for the animals to come back: social impacts of climate changes: a Greenlandic case study', *Journal of Disaster Risk Studies*, vol. 2, no. 3, pp. 235–252.

Wilson, G. and Bryant, R. (1997) *Environmental Management: New Directions for the Twenty-first Century*. UCL Press, London.

Part IV

Arctic energy, investment and legal framework

7   Energy and Arctic investment

Sarah Sternbergh

Introduction

To bring new, sustainable energy solutions to Arctic countries and territories, we will need to consider the balance between ensuring energy security for these territories and their role in leading the way to a future free of fossil fuel sources. Creating an Arctic with energy independence both from fossil fuels and outside imports will reaffirm the northern spirit of entrepreneurship, innovation and will reshape the future of the region. This initiative will be driven and achieved by realizing the region's diverse energy potential, and will be achieved by using new infrastructure platforms and new business models to secure the future sustainability and stability of the Arctic energy management system.

Arctic countries and territories considered in this analysis are those defined according to the administrative boundaries of the Arctic Council, and include: Canada, USA, Greenland (Denmark), Iceland, Norway, Sweden, Finland and Russia.

Existing Arctic energy infrastructure

The primary barrier to energy and infrastructure development in the Arctic is the fact that the majority of the Arctic is comprised of vast undeveloped areas with small, isolated populations. In Canada, Greenland, Russia and Alaska, population density in the Arctic is between 0.03 people/km^2 and 0.46 people/km^2 due to the vast territory and isolated populations.

In much of the Arctic, electrical grid development is limited with many community energy supplies consisting of remote microgrids supplied by fossil fuels—primarily diesel power generation. Heating energy, which represents a significant portion of the energy demand in most of the Arctic, is often supplied to individual buildings from varied heating sources including electrical, diesel burning or wood burning. Highways and other transport grids are limited, and many communities rely on winter roads with very limited opening seasons; sea barges that can only access coastal communities in summer; or, in some cases, solely on air transport. Transport limitations in turn make acquiring construction materials a costly, complicated and energy intensive endeavor in most cases.

The unique environment of the Arctic also presents challenges for infrastructure development such as permafrost, which impedes construction of infrastructure such as buried pipe or electrical lines, highways and building foundations. Furthermore, in many parts of the Arctic, permafrost complicates the heating of buildings, requiring the use of stilts to prevent heat transfer from the building to ground. A limited construction season due to cold winter weather and lack of sunlight can result in major challenges to timely completion of projects. Cold weather and shorter hours of work limit the potential for winter season and extend project schedules. Extreme winter weather and lack of sunlight make implementing wind and solar technology in the Arctic particularly challenging in terms of construction and operation. Additionally, battery technology is negatively impacted by cold temperatures, limiting energy storage solutions.

A challenge that is common to the Arctic and many other remote and undeveloped regions of the world is limited human capacity. Due to remote communities and lack of available education opportunities, human capacity to support technological advancement is restricted. In many communities, students seeking higher education must travel to larger population centers, sometimes requiring them to leave their communities for months or years at a time to obtain training. Once students have finished their studies, there is often little or no opportunity for employment in remote communities for returning graduates and many choose to seek employment in larger population centers rather than return.

Though the above challenges are common in remote Arctic regions, there is significant variability in population density across the Arctic and this translates to variation in levels of development. Based on these variations, the Arctic can be divided into three distinct regions.

Scandinavian Arctic

The Scandinavian Arctic regions of Iceland, Sweden, Norway and Finland have higher population density relative to the rest of the Arctic; ranging from about 2.0 people/km^2 in Finland to about 4.3 people/km^2 in Norway (Statistics Finland, 2016; Statistics Iceland, 2017; Statistics Norway, 2017; Statistics Sweden, 2017). Higher population density translates to large population centers and relatively developed infrastructure including integrated power grids and transport systems such as highways and railroads, which afford Scandinavian Arctic regions unique advantages over the rest of the Arctic in terms of renewable development.

Under the Arctic Council, Iceland is considered to be an Arctic territory in its entirety (Arctic Council, 2018). The population of Iceland is about 334,000 people with a population density of about 3.2 people per square km (Statistics Iceland, 2017). Iceland has a very unique position in the Arctic in terms of renewable energy consumption and production. Electrical and heating energy in Iceland are almost entirely produced by renewables from hydro and geothermal with a small amount of electrical energy produced from wind projects (Iceland National Energy Authority, 2016). In 2015, 97 percent of the electrical generation in Iceland was produced from combined hydro and geothermal and 96 percent of heating power was produced from geothermal resources (Iceland National Energy Authority, 2016). As a result of these developments, 85 percent of Iceland's primary energy consumption was from renewable resources with the remaining 15 percent primarily used in transportation and the fishing industry (Iceland National Energy Authority, 2016).

Swedish Arctic territory consists of Swedish Lapland and, with a population of about 242,000 people, has a population density of about 2.2 people/km² (Statistics Sweden, 2017). Swedish Arctic infrastructure is characterized by relatively high levels of development including integrated power and transportation grids. Much of the energy in Swedish Lapland is supplied by hydro, fossil fuel thermal, biomass thermal and wind sources (Business Index North, 2017).

The Norwegian Arctic consists of part of the county of Nordland, Troms and Finnmark counties and the island of Svalbard. The Arctic portion of Nordland, Troms and Finnmark have a population of 393,000 people and the highest population density of any Arctic region with about 4.1 people/km² (Statistics Norway, 2017). Norwegian Arctic infrastructure is characterized by relatively high levels of development including integrated power and transportation grids. The Norwegian Arctic has significant hydropower development and potential (Business Index North, 2017).

Svalbard is an island archipelago located in the Arctic Ocean and owned by Norway; however, it is a demilitarized area with an international agreement allowing treaty members (Norway, USA, Denmark, France, Italy, Japan, the Netherlands, Great Britain, the British Commonwealth members and Sweden) access to natural resources (The Arctic Governance Project, 2018). Home to 2,210 people in October 2017, Svalbard has a population density of about 0.035 people/km² (Statistics Norway, 2017). The island is the location of important international Arctic research bases and developments including the Svalbard Global Seed Vault. In Longyearbyen, power is supplied from a coal powered plant (Tonseth, 2017).

Finnish Arctic territory is known as Finnish Lapland and is home to a population of about 181,000 people, with a population density of about 1.8 people/km² (Statistics Finland, 2016). Finnish Arctic infrastructure is characterized by relatively high levels of development including integrated power and transportation grids and much of the energy used in Finnish Lapland is supplied from hydropower and biomass (Statistics Finland, 2016).

The Faroe Islands comprise a tiny, isolated Arctic territory in the North Atlantic Ocean governed by Denmark. The population of the Faroe Islands is about 49,000, and, with a population density of about 37 people/km², the Faroe Islands are far more densely populated that other Arctic territories (Statistics Denmark, 2017). However, due to the isolation and extreme weather, the Faroe Islands resemble the small population centers found in the rest of the Arctic and face many of the same challenges in terms of infrastructure development. The Faroe Islands were very dependent on fossil fuels for energy until recent years where they have developed significant renewable energy infrastructure (inFaroe, 2017).

Siberia and Alaska

The Arctic regions of Siberia and Alaska have similar, moderate population density and levels of development. Population densities are about 0.43 people/km^2 in Alaska and about 0.46 people/km^2 in Russia (US Census Bureau, 2017; The Arctic Institute, 2018). Development in both regions is characterized by a few large population centers coupled with tiny remote communities separated by vast wilderness.

Russia has the highest total population residing in the circumpolar Arctic with about two million people living north of the Arctic Circle, and Murmansk, home to about 300,000 people, is the largest city north of the Arctic Circle (The Arctic, 2018). Siberian energy infrastructure outside population centers consists of microgrids that are often reliant on volatile, expensive diesel.

The United States Arctic territory consists of the state of Alaska with a total population of about 800,000 people and two major population centers south of the Arctic Circle in Fairbanks and Anchorage (US Census Bureau, 2017). Energy infrastructure in Alaska, outside of these population centers, is characterized by remote microgrids supplying small communities through diesel power generation.

Greenland and Canadian Arctic territories

Greenland (Denmark) and Canada's Arctic territories have very similar population densities with less than 0.03 people/km^2 (Statistics Canada, 2017; Statistics Denmark, 2017). Greenland and the Canadian Arctic development is characterized by small population centers separated by vast wilderness.

Greenland is home to approximately 56,000 people with a population density of about 0.025 people/km^2 (Statistics Denmark, 2017). Approximately 80 percent of Greenland's landmass is covered by the Greenland ice sheet, and, as a result, Greenlandic communities are isolated along the coast of the country with little or no interconnecting infrastructure. The Greenlandic ice sheet offers significant opportunity for development of hydroelectric power stations and some isolated hydropower plants exist (Government of Greenland, 2018).

The Canadian Arctic consists of three territories; the Yukon, the Northwest Territories and Nunavut. The total population of the three territories is about 111,000 people (Statistics Canada, 2017), with three population centers in Whitehorse, Yellowknife and Iqaluit totaling less than 25,000 people each. Canadian Arctic energy development consists of isolated microgrids with energy demand overwhelmingly supplied by diesel.

Arctic investment channels and current projects

Arctic territories are owned by economically developed countries; however development in the Arctic is limited and much of it consists of small, isolated communities with much lower levels of development than seen in the non-Arctic regions of their respective countries. Investment channels for sustainable energy research and development in the Arctic include initiatives at all levels including international organizations and corporations, state governments, local governments, local corporations, non-governmental organizations and private investors.

Funding for renewable energy development in the Arctic is primarily provided by each regions respective state or federal government. Corporate energy investments in Arctic territories are primarily in the oil and gas industry particularly in the North American, Norwegian and Russian Arctic regions (Holter, 2017; Paraskova, 2017; US EIA, 2017). The major exception to the oil and gas investment trend in private energy investments is seen in Iceland where 85 percent of the energy used in the country is from renewable, low carbon geothermal energy in the form of both direct heat and electrical use and the majority of this energy is provided by privately owned energy companies (Iceland National Energy Authority, 2016).

Government funding sources, policies and projects

Canada

In Canada, the current government policy is supportive of renewable energy development. In real terms, this translates to the federal "Investing in Canada" infrastructure budget with planned funding of $400 million under the Arctic Energy Fund and $9.2 billion for green infrastructure development in Arctic communities over the next 11 years (Government of Canada, 2017). Projects that arise from this focus on development in the Canadian Arctic include geothermal, wind, solar, biomass and hydro power resources as well as energy efficiency initiatives.

Past renewable energy projects funded through previous iterations of federal funding (i.e., Build Canada) include regional level geothermal, wind and solar potential studies completed in Yukon and Northwest Territories; geothermal drilling projects completed in Yukon; wind and solar infrastructure development and energy efficiency considerations for projects built by federal, territorial and first nation governments across Canada's north.

At the territorial (regional) government level, funding for renewable energy infrastructure has long been established. The three territories have individual funding initiatives to support development of renewables.

In Yukon, funding is available from the territorial government through the Yukon Energy Solutions Centre to support energy efficiency in private homes as well as for commercial businesses. The program provides funding to support energy audits; buy-back of electricity produced from renewable resources; rebates to support energy efficiency; low interest loans for retrofitting older buildings with more efficient insulation; funding to assist private homeowners with solar power installation costs; and funding to assess wind energy potential (Yukon Government, 2017). Included in the buy-back program is funding to support power purchase programs for private homeowners and small commercial micro-generation from renewables (Yukon Government, 2017).

In Northwest Territories funding is available from the territorial government through the Arctic Energy Alliance to support energy efficiency initiatives, energy retrofits and power generation from renewables, which are available to various parties including homeowners, businesses, municipalities, non-profit organizations and first nations (Arctic Energy Alliance, 2018). In addition to existing policies, the government of Northwest Territories is currently drafting an energy strategy that includes further renewable energy development (Government of Northwest Territories, 2016). Current and planned renewable energy projects in Northwest Territories include hydro, solar and wind power stations and biomass heating infrastructure.

In Nunavut funding is available from the territorial government to purchase energy generated by renewables from micro-generation and energy efficiency initiatives (Nunavut Energy, 2018). Furthermore, the Nunavut government is currently undertaking renewable energy potential studies including a geothermal potential study and other renewable energy projects at the community level including wind, solar and hydroelectric power generation and geothermal exploration across the region (Nunavut Energy, 2018).

USA

Alaska supports renewable energy projects through the Alaska Energy Authority Renewable Energy Fund (Alaska Energy Authority, 2017). The fund is established and maintained through the Alaska State Legislature and funds renewable energy projects with the intent to reduce and stabilize the cost of energy across the state (Alaska Energy Authority, 2017).

Through this structure financing opportunities are provided by two major programs. The Alaska Housing Finance Renewable Energy Alaska Project has programs that provide low interest loans and energy rebates for improving energy efficiency in homes and businesses (REAP, 2017). Alaska Centre for Energy and Power provides funding for energy research focusing on lowering the cost of energy throughout the state of Alaska. Funding through this program is not only focused on renewables; however, renewable energy solutions including increased efficiency in heating and insulation, energy storage and renewable energy sources such as wind and solar are considered some of the main projects funded through this program (ACEP, 2017).

Current renewable energy projects in Alaska include wind parks, geothermal power generation, and development and research on microgrids aiming to reduce the dependence of remote Arctic communities on expensive fossil fuels transported via air (Alaska Energy Authority, 2017).

Russia

Renewable energy development funding in Russia is provided from the federal government through a variety of funds and initiatives.

The Wind Energy Fund established in 2017 by the Russian Association of the Wind Power Industry (RAWI) and the Russian Investment Industry (RAI) is intended to create investment in wind energy in Russia with guaranteed return on funds such as through a subsidy for a period of seven years. Russia's Arctic renewable energy potential includes high potential for hydro, wind, solar and biomass across the region as well as potential for geothermal development in some areas (Berdin et al., 2017). This funding initiative may serve to drive wind energy development in the Arctic as well as in more populous areas.

In May 2017, Russia launched an energy auction to purchase contracts for 1.9 gigawatts of renewable energy capacity construction (Renewables Now, 2017a) and in June 2017 selected a total of 2.22 gigawatts of renewable capacity construction (Renewables Now, 2017b). While the focus of this was again, not on the Russian Arctic, the result of more renewable energy capacity throughout the country will provide opportunities for renewable energy investment in the Russian Arctic.

Finland

Finland has a long history of investment in renewable energy infrastructure. In 2016, 45 percent of electrical energy in Finland was sourced from renewables and renewable energy investment has long been a major part of the country's budgeting both to support energy independence and combat climate change (Statistics Finland, 2017; RES Legal, 2018).

Funds for development of renewables in Finland's Arctic are provided through the Finnish state government by a variety of mechanisms including feed-in-tariffs for electrical generation from renewable sources and a subsidy for heating in the form of a "heating bonus" to Combined Heat and Power systems (CPH) plants working on biogas and wood fuel. The Finnish government has a new policy under their National Energy and Climate Strategy for 2030 that will change this subsidy and tariff system to a tendering system for new wind power production in 2017 and biogas and wood power plants in 2018 (RES Legal, 2018).

Sweden

Renewable electrical generation in Sweden is promoted by a quota system and funded by the kingdom through a tax regulation and subsidy scheme for renewable electricity produced from biofuels, wind and solar sources. Tax exemptions are used to support renewable heating using biofuels and grants are provided for research and development in wind energy (RES Legal, 2018). As Sweden aims to produce 100 percent of its electricity from renewables by 2040, investment in renewables is a high priority (Swedish Institute, 2018).

A recent project completed in the Swedish Arctic, the Blaiken Wind farm, has a total installed capacity of 247.50 MW (Media Room, 2017).

Norway

Renewable energy in the Kingdom of Norway is supported through a renewable energy quota system using a certificate trading system (RES Legal, 2018).

Norway with its glaciers, mountains and fjords is uniquely suited for hydro power generation, and about 98 percent of electrical power generation in Norway is produced from renewables, most of which is hydropower and the remaining of which is produced from wind.

A key area for development of renewables in Norway is Svalbard. Because the majority of investment in Svalbard is in the form of research grants from the treaty members, the archipelago is a key renewable energy development center in the Arctic. Energy research projects underway on Svalbard are focused on finding alternative energy sources to replace the coal plant at Longyearbyen (Tonseth, 2017).

Iceland

Funding for renewable development in Iceland is primarily supported by private investment with government resources supporting research and development of innovative tools and methods for renewable energy development in the form of subsidies (RES Legal, 2018).

An example of an innovative energy research project in Iceland is the Iceland Deep Drilling Project, which aims to boost the output of geothermal wells and had its most recent drilling project completed in February 2017 (IDDP, 2017).

Denmark (Greenland and Faroe Islands)

Renewable energy funding in Greenland is financed through the Greenland Government Fund for Renewable Energy and Climate (Government of Greenland, 2018). The status of Greenland is currently in flux with Greenland in the process of seeking and gradually obtaining more independence from Denmark. A key impediment to this independence is Greenland's economic status (Mazza, 2015). Like many Arctic territories, the country relies on funding and support from the Danish government; however, renewable energy may provide a key resource to support economic independence from Denmark (Mazza, 2015). Greenland has vast potential for renewable hydroelectric power generation, and the Greenlandic Government is currently investing in developing these resources to attract industrial investment in energy intensive processes such as data servers and aluminum smelting (Government of Greenland, 2018). Other renewable resources in Greenland include significant wind and solar potential that will couple neatly with hydro generation to provide energy storage and reliable electric sources.

The Faroe Islands is highly dependent on outside resources for their energy needs. To combat this dependence, recent government policy in the Faroe Islands developed with the aim of becoming energy independent means that renewables are a key investment area for the Faroese government and investment in renewable energy are made directly through the state owned power company SEV (SEV, 2017). Projects in the Faroe Islands have included hydro and wind power installations as well as energy storage projects (SEV, 2017). Current planning aims to move the Faroe Islands to 100 percent renewable energy by 2030 and projects under consideration include solar installation, wind farms, tidal installations and energy storage projects (SEV, 2017).

Other funding sources and projects

Beyond federal and regional government investment in renewables, other levels of government including indigenous governments (such as Canada's self-governed first nations) and municipal governments are investing in renewables across the Arctic as the impacts of climate change and the need for energy security rise to the forefront of global concern. Focus on projects with renewable and energy efficient infrastructure, as well as developing renewable energy resources at the regional and community level, can be seen across Arctic communities.

Municipal and local government funding and projects

Small municipal governments across the north are choosing to invest in renewables as the challenges of energy security and the looming threat of climate change become more apparent. Renewable energy projects typically address ways to cut dependence on diesel imports, for example the Kluane Wind project currently being undertaken by the small Kluane First Nation government in Canada's Yukon planning to generate 300 kW of wind capacity in order to move the community away from dependence on diesel (Byers, 2017). While the funding for the project is from the Canadian federal and Yukon territorial governments, the impetuous decision to invest available funding in this project is driven by the community (Byers, 2017).

Private funding and projects

Private investment by large corporations in renewable energy resources in the Arctic is limited due to the combined hurdles of lack of economic payoff and the high risks presented by lack of existing development experience, limited transport infrastructure and extreme weather conditions. In the North American, Russian and Norwegian Arctic territories, private energy investment is focused on fossil fuel exploitation far more than renewable development (Holter, 2017; Paraskova, 2017; US EIA, 2017). Private investment in Arctic renewable energy development so far is focused on easy targets such as readily available hydroelectric in Greenland and on areas where government subsidies support infrastructure development and/or renewable power generation.

One way that corporate investment indirectly supports renewable energy infrastructure is in cases where industrial development such as aluminum smelting in Greenland and Iceland; mining projects in northern Canada, Greenland, Russia and the USA; as well as fish farming in Greenland and Iceland provide demand and impetus for the development of renewable resources. If planned and managed appropriately, these opportunities can provide the basis for long term renewable infrastructure development.

At the small business and homeowner level, investment in renewables in the north is becoming more common. Investors, including private small businesses, homeowners, small green tourism companies and development corporations for small first nation governments, are choosing to fund renewable energy projects in many parts of the north as cost-savings measures or as forward looking energy security investments. This level of investment is often tied to the high cost and low reliability of electricity and heating energy in microgrids where users seek to move away from depending on unreliable, high cost energy. In much of the north, these small businesses and private investments are supported by programs at the federal and territorial regional government levels.

Non-government organizations

Renewable energy investment from non-government organizations (NGOs) is primarily focused on research. Organizations that carry out this investment and research include both university and independent NGOs. Funding for NGOs is often provided by government research grants, though private investment and investment from governments outside of the Arctic are possible through these organizations.

Some examples of organizations and universities working on sustainable energy in the Arctic include: the Arctic Institute which invests in research into circumpolar security including energy security (The Arctic Institute, 2018); the Canadian Geothermal Energy Association (CanGEA), which invests in research and development of Canadian geothermal resources; the World Wildlife Fund, which is supporting renewable energy projects in the Arctic through their Arctic Program (WWF, 2017); and The University of the Arctic.

The University of the Arctic is comprised of Arctic member universities from Canada, Denmark, Faroe Islands, Finland, Greenland, Iceland, Norway, Russia, Sweden and the United States and non-Arctic universities and foundations from around the world including China, Korea, Austria, Britain and more.

Through this partnership, several university programs focus on energy development in the Arctic. The Arctic Initiative at Harvard Kennedy School Belfer Center, established in 2017, is focused on issues of sustainability in the Arctic including on Arctic renewable energy development in the fields of science, technology, education and policy (Harvard Kennedy School, 2018). Iceland School of Energy at Reykjavik University is supported by the government and industry in Iceland focusing on science, engineering, policy, education and research relating to sustainable energy development with unique Arctic perspective and strong ties to Iceland's renewable energy industry (Iceland School of Energy, 2017). The Arctic University of Norway established the Arctic Centre for Sustainable Energy in 2017 with the aim of studying the challenges of developing renewable energy in the Arctic and the impacts of greenhouse gases on Arctic communities (The Arctic University of Norway, 2017).

International government bodies

The Arctic Council is an intergovernmental forum comprised of the Arctic countries with territory located in the Arctic and discussed in this chapter: Canada, the Kingdom of Denmark, Finland, Iceland, Norway, the Russian Federation, Sweden and the United States (Arctic Council, 2018). In addition to these states, the Arctic Council has six permanent participant organizations representing indigenous people in the Arctic and observer status is open to non-Arctic States (Arctic Council, 2018). The Arctic Council is funded by the eight member states through a variety of funding mechanisms with individual states funding various individual structures within the council (Arctic Council, 2018).

Six working groups carry out the work of the Arctic Council, and the Sustainable Development Working Group is focused on sustainable development of Arctic communities including infrastructure development and management of natural resources (Arctic Council, 2018). While this group does not focus directly on renewable energy development, the Arctic Energy Summit, hosted in Helsinki in 2017, was funded and organized by the Arctic Council to bring together players in the renewable energy field including industry experts, scientists, energy professionals and policy-makers from around the globe with the focus on energy resources in the Arctic including renewable energy and energy security for Arctic (The Arctic Energy Summit, 2017).

Through the Arctic Council, the Arctic Remote Energy Networks Academy (ARENA) program brings together key stakeholders in remote Arctic renewable energy generation from across the Arctic regions. The initiative brings together key players in the development of these grids across the Arctic to allow for the sharing of strategies and ideas for implementation of renewables in the Arctic (Arctic Council, 2018).

Future Arctic business models for prosperity

Renewable energy development in the Arctic will provide energy security and combat climate change by reducing carbon emissions. For future investments in renewable and sustainable Arctic energy to lead to economic success, input from all the northern stakeholders will be needed. These stakeholders are state, regional and municipal governments, indigenous peoples, research institutions and private industry. Several potential business models will provide opportunity for development and economic success in the renewable energy industry.

Public-private partnerships provide public funds to help reduce investment risk with industry working to develop innovative and efficient energy solutions. Developing energy infrastructure through this model will allow industry to participate under reduced risk condition and provide the opportunity for efficiency through market competition. An example of this type of partnership is Reykjavik University where combined public and industry funding are brought together to support innovation of Arctic renewable development (Iceland School of Energy, 2017).

Private investments in renewable resources through mechanisms such as power purchase agreements will support the development of sustainable electricity generation. These partnerships can benefit both Arctic communities and private companies by providing affordable renewable energy to industry and providing long term sustainable and reliable energy to Arctic communities. An example of this type of partnership can be seen in Greenland where the presence of Alcoa's industrial aluminum smelter has helped drive the development of hydro power resources (Mazza, 2015).

Community level investment in developing sustainable resources can reduce costs and increase energy security. Investment at both the private homeowner level and the municipal level are key to creating sustainable, independent communities. Some very interesting examples of community level investment in renewables are in the field of microgrid management and energy storage. In Alaska, experiments with microgrid metering and energy storage coupled with power generation from wind and solar have allowed communities to reduce diesel fuel consumption and costs by large margins (Klouda, 2017).

Small business models where markets for renewables such as green tourism and farming can allow for economic returns on investment are becoming more common as the market demand for green energy technologies and sustainable energy solutions rises. An example of a business model supported by green tourism is the Chena Hotsprings in Alaska where geothermal power generation attracts tourists from around the world (Arctic Council, 2018).

Summary

The circumpolar Arctic is a complex region with highly variable levels of population density and infrastructure development. In the Scandinavian Arctic there is significant development of infrastructure and renewables whereas in the North American, Russian and Greenlandic Arctic regions development commonly consists of remote isolated communities reliant on expensive energy imports.

Funding for renewables is available through federal budgets across the Arctic through a variety of funding mechanisms and from other important investors including regional and local governments, industry and small businesses, NGOs, educational institutes and private homeowners. Future development of renewable energy resources in the Arctic will provide economic opportunities, increased energy security and reduced energy costs and will allow the Arctic to lead the way in combatting climate change.

References

Alaska Centre for Energy and Power (ACEP), 2017. [Online] Available at: http://acep. uaf.edu/ [Accessed January 12, 2018].

Alaska Energy Authority, 2017. *Renewable Energy Fund.* [Online] Available at: www. akenergyauthority.org/Programs/RenewableEnergyFund [Accessed January 12, 2018].

Arctic Council, 2018. *Arctic Council.* [Online] Available at: http://arctic-council.org/ index.php/en/ [Accessed January 12, 2018].

Arctic Energy Alliance, 2018. *Arctic Energy Alliance – Programs.* [Online] Available at: http://aea.nt.ca/programs [Accessed January 12, 2018].

Berdin, V. K., Kokorin, A. O., Yulkin, G. M. and Yulkin, M. A., 2017. *Renewable Energy in Off-grid Settlements in the Russian Arctic*, Moscow: WWF.

Business Index North, 2017. *Renewable Energy in the North*, s.l.: Business Index North.

Byers, A., 2017. *Kluane Lake Wind Project to Go Ahead, with Investment from Ottawa.* [Online] Available at: www.cbc.ca/news/canada/north/kluane-lake-wind-project-funding-1.4380814.

Government of Canada, 2017. *Budget 2017 Highlights – Indigenous and Northern Investments.* [Online] Available at: www.aadnc-aandc.gc.ca/eng/1490379083439/149037920 8921#nrth [Accessed January 12, 2018].

Government of Greenland, 2018. *Climate Greenland.* [Online] Available at: http://climategreenland.gl/en/ [Accessed January 12, 2018].

Government of Northwest Territories, 2016. *Draft 2030 Energy Strategy – A Path to More Affordable, Secure and Sustainable Energy in the Northwest Territories*, Yellowknife: Government of Northwest Territories.

Harvard Kennedy School, 2018. *Special Initiative – Arctic Initiative.* [Online] Available at: www.belfercenter.org/arctic-initiative/overview-arctic-initiative [Accessed January 12, 2018].

Holter, M., 2017. Statoil Greenlights Key $6 Billion Norway Arctic Oil Project. *Bloomberg*, 5 December, www.bloomberg.com/news/articles/2017-12-05/statoil-greenlights-key-6-billion-arctic-oil-project-off-norway [Accessed January 11, 2018].

Iceland National Energy Authority, 2016. *Hydro Power.* [Online] Available at: www.nea.is/the-national-energy-authority/publications/ [Accessed January 12, 2018].

Iceland School of Energy, 2017. *About Iceland School of Energy.* [Online] Available at: https://en.ru.is/ise/about/about-ise/ [Accessed January 12, 2018].

Iceland Deep Drilling Project (IDDP), 2017. *The Drilling of the Iceland Deep Drilling Project Geothermal Well at Reykjanes Has Been Successfully Completed.* [Online] Available at: https://iddp.is/2017/02/01/the-drilling-of-the-iceland-deep-drilling-project-geothermal-well-at-reykjanes-has-been-successfully-completed-2/ [Accessed January 11, 2018].

inFaroe, 2017. *Faroe Islands Go For 100 Percent Renewable Energy.* [Online] Available at: www.infaroe.com/faroe-islands-go-for-100-percent-renewable-energy/ [Accessed January 12, 2018].

Klouda, N., 2017. Energy Dept. Grant Aims to Harden Microgrids. *Alaska Journal of Commerce*, October 4.

Mazza, M., 2015. The Prospects of Independence for Greenland, between Energy Resources and the Rights of Indigenous Peoples (with Some Comparative Remarks on Nunavut, Canada). *Beijing Law Review*, 6(doi: 10.4236/blr.2015.64028.), pp. 320–330.

Media Room, 2017. *Blaiken – Groundbreaking Arctic Wind Farm Project Inaugurated.* [Online] Available at: www3.fortum.com/media/2017/09/blaiken-groundbreaking-arctic-wind-farm-project-inaugurated [Accessed January 11, 2018].

Nunavut Energy, 2018. *Nunavut Energy.* [Online] Available at: www.nunavutenergy.ca/ [Accessed January 13, 2018].

Paraskova, T., 2017. Russia Goes All in on Arctic Oil Development. *USA Today*, 24 October, www.usatoday.com/story/money/energy/2017/10/24/russia-goes-all-arctic-oil-development/792990001/ [Accessed January 13, 2018].

Renewable Energy Alaska Project (REAP), 2017. *Renewable Energy Alaska Project.* [Online] Available at: http://alaskarenewableenergy.org/ [Accessed January 12, 2018].

Renewables Now, 2017a. *Russia Selects 2,221 MW of Projects in Renewable Auction.* [Online] Available at: https://renewablesnow.com/news/russia-selects-2221-mw-of-projects-in-renewable-auction-572327/ [Accessed January 12, 2018].

Renewables Now, 2017b. *Wind Investment Fund of US$1.7bn to Be Launched in Russia.* [Online] Available at: https://renewablesnow.com/news/wind-investment-fund-of-usd-17bn-to-be-launched-in-russia-576346/ [Accessed January 12, 2018].

RES Legal, 2018. *Legal Sources on Renewable Energy – Search by Country.* [Online] Available at: www.res-legal.eu/search-by-country/ [Accessed January 13, 2018].

SEV, 2017. *Tangible Plan for the Green Course.* [Online] Available at: www.sev.fo/Default.aspx?ID=193&Action=1&NewsId=2921&M=NewsV2&PID=392 [Accessed January 12, 2018].

Statistics Canada, 2017. *Estimates of Population, Canada, Provinces and Territories, Quarterly (Persons).* [Online] Available at: www5.statcan.gc.ca/cansim/a26?lang=eng&retrLang=eng&id=0510005&&pattern=&stByVal=1&p1=1&p2=31&tabMode=dataTable&csid= [Accessed January 12, 2018].

Statistics Denmark, 2017. *Statistical Yearbook 2017 Faroe Islands and Greenland.* [Online] Available at: www.dst.dk/Site/Dst/Udgivelser/GetPubFile.aspx?id=22257&sid=faroe [Accessed January 12, 2018].

Statistics Finland, 2016. *Area, Population and GDP by Region.* [Online] Available at: https://stat.fi/tup/suoluk/suoluk_vaesto_en.html [Accessed January 12, 2018].

Statistics Finland, 2017. *Renewable Energy Sources Produced 45 per cent of Electricity and 57 per cent of Heat.* [Online] Available at: www.stat.fi/til/salatuo/2016/salatuo_2016_2017-11-02_tie_001_en.html [Accessed January 12, 2018].

Statistics Iceland, 2017. *Population Development 2016.* [Online] Available at: www.statice.is/publications/publication-detail?id=58377 [Accessed January 12, 2018].

Statistics Norway, 2017. *Statistics Norway.* [Online] Available at: www.ssb.no/en/ [Accessed January 13, 2018].

Statistics Sweden, 2017. *Population Density per Sq. km, Population and Land Area by Region and Sex. Year 1991–2016.* [Online] Available at: www.statistikdatabasen.scb.se/pxweb/en/ssd/START__BE__BE0101__BE0101C/BefArealTathetKon/?rxid=4d9f2de9-2736-4602-a516-09ba522519a4 [Accessed January 12, 2018].

Swedish Institute, 2018. *Sweden Tackles Climate Change.* [Online] Available at: https://sweden.se/nature/sweden-tackles-climate-change/ [Accessed January 13, 2018].

The Arctic, 2018. *Population.* [Online] Available at: http://arctic.ru/population/ [Accessed January 12, 2018].

The Arctic Energy Summit, 2017. *About – Organization of the Arctic Energy Summit.* [Online] Available at: http://arcticenergysummit.com/story/About [Accessed January 12, 2018].

The Arctic Governance Project, 2018. *The Svalbard regime.* [Online] Available at: www.arcticgovernance.org/the-svalbard-regime.4668236-137746.html [Accessed January 12, 2018].

The Arctic Institute, 2018. *The Arctic Institute Centre for Circumpolar Security Studies.* [Online] Available at: www.thearcticinstitute.org/ [Accessed January 12, 2018].

The Arctic University of Norway, 2017. *Arctic Centre for Sustainable Energy – ARC.* [Online] Available at: https://en.uit.no/forskning/forskningsgrupper/gruppe?p_document_id=453700 [Accessed January 12, 2018].

Tonseth, S., 2017. *Svalbard's Electric Power.* [Online] Available at: www.sintef.no/en/latest-news/svalbards-electric-power-could-come-from-hydrogen/ [Accessed January 11, 2018].

US Census Bureau, 2017. *Quickfacts Alaska.* [Online] Available at: www.census.gov/quickfacts/AK [Accessed January 12, 2018].

US Energy Information Authority (US EIA), 2017. *Alaska State Profile and Energy Estimates.* [Online] Available at: www.eia.gov/state/analysis.php?sid=AK [Accessed January 12, 2018].

World Wildlife Fund (WWF), 2017. *Fueling Change in the Arctic – Phase II.* [Online] Available at: http://assets.wwf.ca/downloads/full_report___feasibility.pdf [Accessed January 12, 2018].

Yukon Government, 2017. *Energy Mines and Resources Programs.* [Online] Available at: www.energy.gov.yk.ca/programs.html [Accessed January 12, 2018].

8 Corporations and the renewable energy transition in the Arctic

Dyveke F. Elset

Introduction

Climate change is changing corporations' energy procurement practices, and ultimately the global energy system. Since 2012 there has been a drastic increase in multinational corporations (from now on referred to as corporations) actively procuring renewable energy for their services, effectively greening the electrical grid[1] (Business Renewables Center, 2018). This chapter seeks to unravel key mechanisms that are contributing to this central aspect of the renewable energy transition, and investigate how corporations can play a role in the Arctic.

This chapter suggests that there is a potential for corporations within relevant sectors to contribute to the renewable energy transition in the Arctic. However, recognizing the central challenges in the Arctic region such as sparse population, lack of infrastructure and harsh climate, it argues that such corporate impact is most likely to happen in the industrialized or industrializing areas of the region.

This chapter is divided in two: the first part sets out to investigate mechanisms of energy transitions and corporations' role in this on a global level; the second part applies these findings to the Arctic region.

Energy transitions

Energy is complex. It infiltrates every aspect of all social and economic activity: the food we consume, the fuel we tank our cars with, our sources of electricity and heat, and consequently the quality of air in our lungs and the temperature of our planet.

There are several definitions of energy transitions. It can be understood as shifts in fuels and corresponding technologies that allows for the exploitation of these fuels. It can also be understood as a change of energy consumption patterns in a society. A broader understanding is the event in which an economic system relying on one energy source moves to relying on another (Sovacool, 2016, p. 203).

However, energy scholars are skeptical as to whether a successful renewable energy transition will materialize. Vaclav Smil describes the energy regime as "the most expensive anthropogenic infrastructure that cannot be either written-off or displaced rapidly" (Smil, 2016, pp. 194–196). Moreover, Pulitzer Prize winner and energy specialist, Daniel Yergin, argues that previous energy transitions were the result of a "gradual development and adaptation of new technologies" competitive in price and usefulness (Yergin, 2013). The assumptions that the energy system cannot be intentionally changed, and that its evolvement is extremely slow and path dependent, have led scholars to compare it with former energy transitions that have been unintentional and gradual (Grubler, 2012; Fouquet, 2010).

Such reasoning is built on the assumption that the renewable energy transition will happen within the same context and rationale as previous energy transitions. However, as Robert Allen points out, the renewable energy transition differs as it is faced with the "externality issues raised by global warming." The future, he therefore argues, "will not – and should not! – be a replay of the past!" (Allen, 2012, p. 17). For this reason, the focus in this chapter is on observed changes in the context in which energy exist.

The scope of the research is limited to the electricity sector as this is where the renewable energy transition is best observed, and a field with large potential to cut CO_2-emissions (IRENA, 2017). Figure 8.1 gives an overview of world electricity production by source. It clearly shows that while coal is in decline and renewables are growing, a significant effort is needed to increase their share in the energy mix used to produce electricity for it to classify as a renewable energy transition. This forms the background of which we will seek to understand

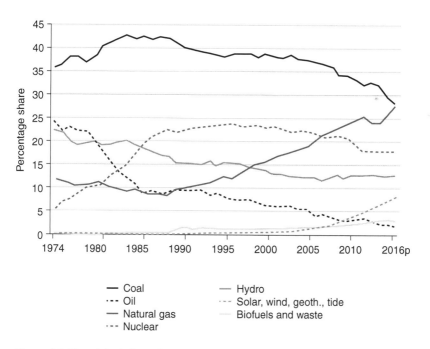

Figure 8.1 Electricity information.

Source: IEA, 2017 (used with permission from OECD/IEA).

corporations' role in the greening of the grid, and ultimately, in the renewable energy transition.

Corporations' responses to climate change

Corporate sustainable procurement is to a large extent a twenty-first century phenomenon, and academic understanding of this is still at an early stage. Yet, existing literature theorizes that corporations' engagement with social and environmental challenges is more likely to be caused by external pressure than internal pressure (Campbell, 2007; Pintea et al., 2014; Rugman and Verbeke, 1998). John Campbell finds that:

> corporations act in socially responsible ways the more they encounter strong state regulation, collective industrial self-regulation, NGOs and other independent organizations that monitor them, and a normative institutional environment that encourages socially responsible behaviour.
>
> (Campbell, 2007, p. 962)

Supporting this argument, Pintea et al. (2014) observe a correlation between the level of a corporation's environmental performances and the development stage of the country in which it operates. Moreover, Alan Rugman and Alain Verbeke argue that corporations "design their production processes according to best global practices" in order to be congruent "with the most stringent environmental regulations prevailing in the relevant countries where they operate" (Rugman and Verbeke, 1998, p. 372). This is also consistent with David Levy's findings that international pressure forces corporations' environmental standards up (in Rugman and Verbeke, 1998, p. 372). Such safeguarding strategies are also found in Adam Bumpus' research, who argues that "firms still focus more on navigating the politico-economic system rather than embracing actions that can be described as a radical departure from business as usual" (Bumpus, 2015, p. 486).

Despite empirical evidence suggesting that climate change "deeply affects purchasing and supply-management practices" (Crespin-Mazet and Dontenwill, 2012, p. 207), companies' procurement engagement itself is often overlooked in academic literature as it is merely viewed as a reactive response to external change (Hoejmose and Adrien-Kirby, 2012, p. 236).

However, there are several examples suggesting that private corporations are taking on a more political role, and that they are perceived as influential international actors. For example, US companies are now pushing climate action by committing to their sustainability goals in the wake of the less climate-friendly political course set by the current Trump Administration (The Energy Gang, 2016).

Moreover, corporation's procurement strategy is an essential tool to reach its sustainability objectives (Crespin-Mazet and Dontenwill, 2012, p. 207), which reveals "a key financial mechanism available to governments to drive policy change" (Correia et al., 2013). Corporations' procurement strategies should, therefore, arguably receive more attention both in academia and by governments. Before analyzing corporations' role in the renewable energy transition, it is necessary to conceptualize energy and energy systems.

Energy as a socio-technical system

In its simplest terms, energy can be defined as the system that converts "energy fuels and carriers" into the energy cycle (Goldthau and Sovacool, 2012, p. 233). However, as established earlier, energy has distinctive social and economic aspects to it as well. The link between technology and social values in the energy system is evident, and it is thus often characterized as a *socio-technical system* (Fouquet, 2010; Sovacool, 2016). Socio-technical systems are defined as "shared cognitive routines" that "contribute to patterning of technological development" (Geels and Schot, 2007, p. 400), and are embedded in institutions and infrastructures in a society (Rip and Kemp, 1998, p. 338).

How then can we better understand the renewable energy transition by thinking about it as a socio-technical transition? The multi-level perspective provides an analytical framework that can help us identify the mechanisms involved in the renewable energy transition.

The multi-level perspective

The Multi-Level Perspective model (MLP) has traditionally been used to construe how new technologies emerge and change systems (Shove and Walker, 2007, p. 764). Here, such transitions occur due to developments on three levels where different actors operate. It suggests that, at the micro-level, individual actors such as companies and environmental movements operate. At the meso-level, networks, communities and organizations operate. And at the macro-level, nations and federations of states operate (Rotmans et al., 2001, p. 19). Developments on each of the levels could alter the system by affecting behavior on other levels, consequently causing a system transition.

Frank Geels and Johan Schot propose a pathway-approach to understand socio-technical system transitions. Here, the timing and nature of the multi-level interaction best explains how socio-technical systems change and the degree of the actors' involvement in this change (Geels and Schot, 2007, pp. 407–408).

The Transformation Path, visualized in Figure 8.2, best explains the way in which the renewable energy transition occurs. Here, there is a pressure at the macro-level to change (increased concern about climate change) but at a time where the technology (renewable energy capacity and conveyors) needed is yet to be fully developed. Consequently, system actors seek to respond by "modifying the direction of development paths" (Geels and Schot, 2007, p. 408). Essential here is that the transition only materializes if the changes on the macro-level is "perceived and acted upon" by the actors in the system (Geels and Schot, 2007, p. 406).

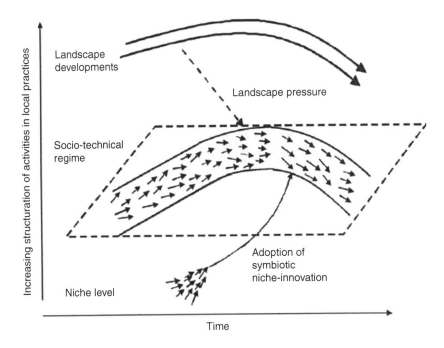

Figure 8.2 Transformation pathway.

Source: Geels and Schot, 2007, p. 407 (used with permission from Elsevier).

The MLP only includes nation states and international governmental organi-zations in the analyses of the formation of institutional pressure at the macro-level (Rotmans et al., 2001, p. 19). However, Kenneth W. Abbot and Duncan Snidal argue that the governance of the macro-level and meso-level is altering as non-state actors are altering business norms on an international level (Abbot and Snidal, 2009). Here, corporations will be included in the analysis of the macro-level as this would arguably give a more accurate portrayal of the topic at hand.

Case study: Google

Google is one of the most powerful and influential multinational corporations in the world. In 2016, it received the *Diplomat of the Year Award* from Foreign Policy for profoundly changing the nature of international relations, human rela-tions and the way we see the world (Foreign Policy, 2016). Moreover, Google is the biggest procurer of renewable energy in the world, which makes it a suitable case study for the question asked here: what is the role of corporations in the renewable energy transition? Following the structure of the Transformation Pathway, the following sections will investigate Google's involvement on the different levels in the MLP model.

Google's involvement at the macro-level

The United Nations Framework Convention on Climate Change (UNFCCC) clearly established the urgent need to act on climate change in the Intergovernmental Panel on Climate Change (IPCC) report in 2014. The Paris Agreement (UN, 2015) reached the following year indicates that the need to stop climate change has reached an international consensus, which is arguably a turning point that has gathered a momentum for efforts seeking to fight climate change.

The need for a renewable energy transition has become one of twenty-first century's biggest challenges occupying actors on the macro-level. The Clean Energy Ministerial (CEM) was established at UNFCCC COP19 in 2009 to serve as a high-level forum where key public and private actors can develop and promote programs and policies that enhances the renewable energy transition (CEM, 2017a). The CEM differs from the UNFCCC Conference of the Parties (COP) as it encourages the private sector's involvement and input in policy formulation (CEM, 2017b). Its member states account for 75 percent of world emission and 90 percent of global clean energy investment (CEM, 2017a), and Google's involvement here is therefore central to the questions at hand.

Especially important is the Corporate Sourcing of Renewables Campaign launched at the CEM's seventh meeting in 2016. The goal is to "get more companies to commit to powering their operations with renewables and deploy tools and resources to enable more companies, large and small, to do so" (CEM, 2016a). The partners include the several governments, together with the International Renewable Energy Agency (IRENA), The Renewable Energy Buyers Alliance (REBA), RE100, World Business Council on Sustainable Development (WBCSD), the World Resource Institute (WRI). Google has together with several other companies[2] officially committed to this campaign (CEM, 2016a). They state that they will lead by example and provide "models for procuring renewables" and "collaboratively advocate for supporting policies" in order to overcome the existing barriers of corporations' procurement of renewable energy. This includes implementing and developing reporting mechanisms that include energy consumption and emission, and simplifying Power Purchasing Agreements (PPAs), which will allow for smaller corporations to follow suit (CEM, 2016b).

As a direct result of the CEM initiatives, more countries have now adopted or proposed standards and policies that will improve their energy efficiency, and implement internationally recognized Energy Management Systems within their jurisdictions (CEM, 2016a). The CEM's efforts are clearly manifesting renewable energy as a solution to climate change. Indeed, Energy Secretary under the Obama Administration, Ernest Moniz, stated at the seventh CEM in San Francisco in June 2016:

> The discussion about climate and climate science and mitigation and adaptation is very important, but the fact is that clean energy, the clean energy scaling, has an inevitability about it following Paris … This is the direction we're going.
>
> (McMahon, 2016)

Google's contribution to the CEM has been in efforts together with the above listed participants. They have committed to:

> Work in collaboration with national governments, renewable energy buyers and suppliers, NGOs, and others across CEM member governments to convene a group of stakeholders to survey the policy and regulatory structures critical to enabling corporate purchasing of renewable energy and make policy recommendations for how to further unlock corporate renewable energy demand in CEM member governments.
>
> (CEM, 2016c)

By sharing their experiences (Kimball, 2016) and helping to identify barriers and opportunities to corporate procurement of renewable energy, Google has essentially served as a "corporate champion" amplifying "the corporate buyers' voice" (CEM, 2016b).

On the macro-level, this forum has established that climate change mitigation is of pivotal importance, that a renewable energy transition is needed and that enhancing corporate procurement of renewables is an important tool in order to succeed. This clear positioning on the issue forms "the landscape pressure" illustrated in Figure 8.2. It further shows that the private sector has been instrumental in the formation of this narrative, and it specifically shows how Google has contributed to this development. How the transition will occur is determined by how niche-innovations at the micro-level respond to the landscape pressure at the macro-level, and how these innovations will be adopted by actors on the meso-level and ultimately become mainstream.

Developments at the micro-level

At the micro-level, individual actors operate, innovate and adjust to new circumstances, which could ultimately bring about changes on the meso-level. This section will outline key technological, business and financial innovations that are facilitating the renewable energy transition.

Modern renewable energy technology

Power generation is the end-use sector that has experienced the most rapid growth of renewable energy adoption (IRENA, 2017, p. 64). It therefore serves as a suitable focal point of researching energy transition, as it can give insight into how niche technologies become mainstream. Solar photovoltaic (PV) and wind power are the two sources of renewable energy that have had the highest capacity growth rate of scale between 2010 and 2015. Solar PV saw a 41.4 percent increase, and onshore and offshore wind saw a 17.7 percent increase and 30.1 percent increase respectively (IRENA, 2017, p. 68). Some key innovations in these sectors are highlighted below.

Solar PV is growing due to its modular design, which offers a variety of installation options. Solar modules can provide on-grid solutions, utility-scale projects or off-grid modules (IRENA, 2017, p. 69). Supportive policies and falling costs further leads to technological progress and reduced prices (IRENA, 2017, pp. 69–70). One essential innovation is the non-silicon technologies, which are more efficient than previous films (IRENA, 2017, p. 70). Another, more notable innovation, are the more fashionable designs of rooftop solar PV modules that looks like regular rooftop tiles. These are "thin, photovoltaic (PV) sheets that can replace or lay on top of existing" tiles in any climate or environment, effectively capturing solar energy (Solar Power Authority, 2016). One can therefore argue that solar PV technology emerges as a flexible, user friendly and increasingly efficient energy alternative to the traditional energy sources.

Despite these developments, only 1.2 percent of the global electricity output in 2015 was generated from solar PV. This was mainly due to its low base point for growth compared to conventional technologies; its still relatively low capacity factor; and finally, due to an overall rise in demand for electricity (IRENA, 2017, p. 70). This suggests that solar PV should still be considered a niche-technology.

As for wind power, onshore wind is generally considered a mature technology. The technological innovation in the wind industry is currently happening offshore—opening doors to more potent areas (IRENA, 2016, p. 1). The last decades' technological innovations have increased the capacity of individual turbines (from 2 MW to more than 6 MW) and reduced the costs of offshore wind (IRENA, 2016, p. 4). Moreover, developments in "foundations, installation, access, operation and system integration" have enhanced wind power capacity (IRENA, 2016, p. 2). Furthermore, advancing experimental work on floating foundations could make the installation process more economical. Lastly, innovation in underwater electrical interconnections, such as the high-voltage direct-current (HVDC) cable now used, could drive the costs of the transmission down and pave the way for offshore international super-grids (IRENA, 2016, p. 5).

Since 2002, the installed offshore wind capacity has grown from 160 MW to 12 GW globally (IRENA, 2016, p. 2). Wind power alone has "avoided over 637 million tonnes of CO_2 emissions globally in 2015" according to the Global Wind Energy Council (GWEC), which is twice as much as the solar PV sector. Yet wind was only responsible for 3.7 percent of the global electricity generation in 2015 (GWEC, N/A). These developments suggest that wind technology is more developed than solar PV, but that there is further potential for it to become more efficient and widespread.

Technological innovations in niche markets, such as the solar PV and wind industries, are driving down the costs and becoming more available to the end power consumers, however, they remain dependent on support from regulators and investors (IRENA, 2017). One of the companies that has invested in these technologies is Google. Google installed a 1.6 MW solar PV panel on the Google plex in California, already back in 2006. In the press release, Google stated that this was their "first step" in reducing their "environmental impact as a company" (Beavers, 2006). This is important because Google's adoption of new renewable energy technologies happened at a time when these technologies were more costly, and a climate change consensus was yet to be reached.

Google is currently the world's largest corporate buyer of renewable energy and in December 2017 they announced that it has become 100 percent powered by renewable energy (Donnelly, 2017). This is a result of a long-term strategy to become carbon-neutral, where renewable energy procurement has been "one of the most important tools" (CDP, 2017, pp. 2, 3). The next section will therefore look into renewable PPAs.

PPAs

PPAs are contracts between electricity generators and consumers (Dunlop and Roesch, 2016). Traditional PPAs provide power from existing energy sources, such as coal, gas and oil, and are often operations-based, dependent on location and short termed. Renewable PPAs, on the other hand, refer to the sourcing of energy from renewable energy projects yet to be built. By signing renewable PPAs corporations are effectively agreeing to finance the expansion of renewable energy (Royal, 2016).

Google has signed 20 PPAs around the world since 2006, which accounts for more than 2.6 GW of renewable energy (Google, 2017, p. 12). Its PPAs strive to meet strict criteria that will (1) "create new sources of green power on the grid"; (2) apply the renewable energy to Google's power consumption the same year the energy is generated and; (3) ensure the same amount of green attributes[3] and bundled energy[4] in each purchase (CDP, 2017, p. 7). Google argues:

> By being an early investor and deploying smart capital to fund utility-scale projects, we believe we can accelerate the deployment of the latest clean energy technologies while providing attractive returns to Google as well as more capital for developers to build additional projects.
>
> (CDP, 2017, p. 20)

There are several examples of renewable energy deployment projects where Google has been involved, such as the Tellenes wind farm-deal struck between Norwegian Wind Energy, Zephyr and Google in June 2017 (Norwea, 2016). It also worked directly with utility suppliers in Iowa and Oklahoma to develop wind energy to power their operations (Sotos, 2014, p. 12). Moreover, Google recently committed to "invest in the largest wind farm in Africa, which, when completed, will have a capacity of ~15% of the current grid in Kenya" (CDP, 2017, p. 20).

These PPAs with Google are important. By providing utilities with a guaranteed revenue stream for a long period, Google helps to reduce the risk of producing renewable energy (Google, 2013, p. 4). This is effectively contributing to energy security, supply and greening of the grid.

Changes at the meso-level

Having outlined the developments at both the micro and macro-level, this section will map out how these have materialized on the meso-level so far. In Figure 8.2, this is called the "socio-technical regime," which is where we find evidences of a system transitioning. Below, changes in business and finance practices are used as indications of a renewable energy transition.

Business practices and the procurement of energy

There is a myriad of businesses, business networks and collaborations, and elaborating on each of them is beyond the scope of this chapter. It will instead investigate Google's involvement with one of these communities: Renewable Energy 100 (RE100). RE100 is "a collaborative, global initiative" of 87 businesses "committed to 100% renewable electricity" (RE100, 2017, p. 3). The initiative encourages companies to set emission reduction targets and renewable energy procurement targets, and to share best practices and knowledge to "strengthening the market signal needed to transition the global economy away from fossil fuels and towards net-zero emissions" (RE100, 2017, p. 5). In its annual report of 2016, it states:

> The continued growth in membership following COP21 shows that RE100 is the essential global initiative through which corporates can demonstrate their climate leadership, seize commercial advantage, and develop the markets that will create a tipping point in the transition to renewable energy.
> (RE100, 2017, p. 1)

Through its Technical Advisory Group, the RE100 has published guidance on how to help its members procure renewable energy (RE100, 2017, p. 20). It reported that its members have cumulatively reached 22 percent of their set target in 2015, but it did not provide data on their total procurement per 2017 (RE100, 2017, p. 5). It is therefore not possible to determine if this campaign has had an actual effect on the amount of renewable energy procured by the RE100 members. However, communicating the business case for renewable energy procurement by sharing the success stories from the signatories is arguably encouraging this procurement model.

RE100 is cooperating with the Rocky Mountain Institute Business Renewable Center in bringing about business-to-business opportunities by pairing sellers and buyers. Together they have hosted webinars on topics including PPAs (RE100, 2017, p. 20). RE100 is also cooperating with IRENA, WBCSD, REBA and ShareAction[5] in order to influence governments, investors and other companies to enhance the uptake of renewable energy (RE100, 2017, p. 21).

By organizing and participating in key energy and climate change events in the US, Europe, India and China, RE100 is displaying their members' commitment to renewable energy. Examples are COP21 and 22, the Business & Climate Summit, Climate Week NYC and the seventh Clean Energy Ministerial. This is allowing its members "to demonstrate their climate leadership to key audiences" (RE100, 2017, p. 21).

While the RE100 is helping their members, the organization is also depending on reliable and well-established companies to make their message trustworthy. Google, being one of the biggest companies in the world, is therefore serving as an important example of RE100's work. In Google's work with RE100, they have communicated that increased renewable energy procurement makes business sense. Senior Vice President for Technical Infrastructure at Google, Urs Hölzle, stated in RE100's Annual Report:

> Electricity costs are one of the largest components of our operating expenses at our data centers, and having a long-term stable cost of renewable power provides protection against price swings in energy.
>
> (Google, 2017, p. 8)

Google made their PPA strategies available already in 2013, which contributed to knowledge-sharing among corporations and showed a clear attempt to influence other companies' energy procurement practices (Google, 2013).

The purpose with this section was to demonstrate how business networks, such as the RE100, are collaborating in order to bring about changes in business practices and to convince regulators that facilitating procurement of renewable energy make business sense. Former Managing Director of the Rocky Mountain Institute and co-manager of the Business Renewable Centre, Hervé Touati, called the alliance built among corporations working to enhance corporate procurement a "movement" (The Energy Gang, 2016), suggesting that this indeed is an intentionally driven change in dominant corporate practices and assumptions. Moreover, the observations here suggest that businesses conform to the threat of climate change by framing the procurement of renewable energy as a good business case, making it easier to incorporate the overarching issue of climate change into their business practices.

Financing practices and the procurement of energy

New forms of financing renewable energy projects are emerging and confirming the renewable energy transition (IEA, 2016). Bloomberg New Energy Finance reports that investment in renewable energy fell for the first time in 2016, yet the number of projects continued to rise due to lower costs. Moreover, they observe that investments are now going toward grid capacity and infrastructure to digest the added renewable energy. Acquisitions and corporate merging activities into renewable energy projects have also spurred the last couple of years (BNEF, 2017). Moreover, a recent report by Moody's Investors Service concluded that "contracts to sell electricity directly to corporate users are among the key demand drivers" for renewable energy (Eckhouse, 2017).

The Institute for Energy Economics and Financial Analysis finds that in more developed markets new yieldcos,[6] infrastructure funds and green bonds are steering capital into renewable energy projects. In developing markets, where the risks are too high to use these types of financing mechanisms, alternative models are developed to meet the demand for capital for renewable energy projects. Here, multinational development banks are acting as catalysts by "helping companies to get access to capital markets and to proceed with plans they might otherwise abandon" (Buckley, 2016). Another financing innovation is the coordinated financing between government agencies, development banks, states and companies. Moreover, as pension funds, sovereign wealth funds and insurers are perceiving renewable energy projects as good investments, "entire bond and equity markets" are now evolving as institutional investors in renewable energy (Buckley, 2016).

However, procuring renewable energy is not only an issue of costs. Greentech Media (GTM) Research solar analyst, Colin Smith, explains "you're talking about very complex deals and arrangements with unique risk profiles that most companies aren't fully well equipped to understand" (Moodie, 2016). Big companies such as Google have "the resources to understand the energy markets and negotiate contracts to buy renewable energy" (Moodie, 2016), which therefore plays an important role on a meso-level.

Corporations' role in the renewable energy transition

By using the Transformational Pathway as an analytical tool, this case study of Google has given us an understanding of how societal and technical developments form the renewable energy transition, and particularized the role of corporations in this process. It finds that the relationship between climate change and renewable energy is considered a moderate landscape pressure formed at the macro-level. On the micro-level, solar PV and wind technologies as well as new business strategies are key niche-innovations that provide alternatives to business as usual. At the meso-level, changes of financial and business practices geared toward accommodating the developments on the micro and macro-levels suggests that a renewable energy transition is indeed underway.

The findings furthermore suggest that corporations are influential on all levels (micro, meso and macro). On a macro-level, corporations' preferences influence policymakers, which contribute to the infrastructure needed to procure renewable energy. On the micro-level, corporations innovate their strategies to adapt to new technologies, such as seen with PPAs. Google exemplified that being an early procurer of renewable energy means not only conforming to existing norms and practices, but also forming them. Finally, the study shows that individual companies actively promote the procurement of renewable energy through PPAs to other potential buyers and providers.

The Transformation Pathway tells us that for a transition to occur, the desired changes formed at the macro-level must be understood and acted upon by other actors on the meso and macro-level. However, it is unable to explain why actors conform to these new norms. The next section will show how the Logic of Appropriateness (LoA), an institutional theory perspective first used by James

March and Johan Olsen (1988), can shed light on why corporations are conforming to the sustainable procurement of energy, which is ultimately contributing to the renewable energy transition.

The Logic of Appropriateness in the renewable energy transition

The LoA suggests that norms and rules in a society are formed according to what is viewed as appropriate. Martha Finnemore (1996) explains it as follows:

> Norms of behaviour and social institutions can provide … direction and goals for action. The value they embody and the rules and roles they define channel behaviour. Actors conform to them in part for "rational" reasons but also because they become socialized to accept these values, rules, and roles. They internalise the roles and rules as scripts to which they conform, not out of conscious choice, but because they understand these behaviours to be appropriate.
>
> (Finnemore, 1996, p. 29)

The developments on the macro-level suggest that the need to combat climate change has manifested itself as a war-like situation in which everyone needs to contribute. In other words, climate mitigation is now considered the right thing to do. The mission of the CEM further establishes the central role renewable energy has in this battle. This clear positioning serves as guidelines for businesses and other actors operating in the energy sector.

The innovations at the micro-level regarding financing practices and PPAs facilitates the renewable energy transition, but only if they become mainstream practices. The work at the meso-level, where there is an inclination to frame climate change as a business opportunity, is therefore important. The purpose of the RE100 campaign is to help firms showcase their climate leadership, clasp the commercial advantage and consequently forge a tipping point in the renewable energy transition. The underlying argument is that the transition will happen regardless, and that a low-carbon footprint is considered a competitive advantage.

Finnemore finds in her research that there is usually a mix of logics playing in norms and rules of governance. In analyzing states' behavior and interests, she finds that "agents create social structure for consequentialist reasons but they spread for reasons of appropriateness" (Finnemore, 1996, p. 30). The logic of consequence is here referred to as the idea that "pre-specified actors or agents make means-ends calculations and devise strategies to maximize utilities" (Finnemore, 1996, p. 30). Interestingly, Google's analysis suggests that the reverse is happening with the corporations involved in the renewable energy transition: the social structures are created for reasons of appropriateness (i.e., combatting climate change), but spread for consequentialists reasons (i.e., to unlock the business potential climate change presents).

So far, this chapter has revealed some of the key mechanisms at play in the renewable energy transition. International cooperation between governments, businesses and organizations has paved the way toward a renewable future. It has argued that renewable energy related innovations are utilized and developed by large corporations, due to a LoA, and spread due to a strong business case being built around these models. This could provide some useful indications as to how a renewable energy transition could emerge in the Arctic.

Corporations and the renewable energy transition in the Arctic

Corporations could be a vehicle for the development and exploitation of renewable energy in the Arctic. However, vast and sparsely populated areas with harsh weather conditions are making the region less attractive to corporations (Hope et al., 2015). Corporate involvement with renewable energy development, as discussed in this chapter, is therefore most likely to occur in industries already operating in the Arctic.

There is a supply and demand for renewable energy in the Arctic. The main industries in the region are energy intensive; mineral extraction, fishery, farming, shipping, and oil and gas (Duhaime and Caron, 2017, p. 17), and the region has an abundance of renewable energy sources. Wind and hydro-power are the most potent ones, and Iceland is making a strong case for geothermal power. One could therefore argue that there is a vast potential for corporate procurement of renewable energy in the region. However, fossil fuels remain the main source of electricity generation in several parts of the Arctic (Nordregio, 2011). Many remote areas in the region are still off-grid and rely on long-traveled diesel for lightning and heating. For the purpose at hand, it is therefore important to make the distinction between the off-grid and the on-grid Arctic as they face different challenges in the renewable energy transition.

In off-grid areas with not enough community capacity to operate an energy intensive business, private corporations will have few incentives to move their operations there and develop renewable energy installations. These areas will need alternative solutions for renewable energy development,[7] which falls outside the scope of this chapter. The potential for corporate involvement as discussed here thus lies in the on-grid areas of the Arctic.

Renewable energy development is not getting enough attention at the macro-level in the Arctic region. The circumpolar states differ in their communication of renewable energy development in the region. On one hand, the Norwegian government has announced that they will promote access to renewable energy resources in the Norwegian Arctic as "a basis for business development and value creation" (Norwegian government, 2017, p. 32). On the other hand, in the significantly larger and less developed Russian Arctic, renewable energy development is not seen as a primary focus (Berdin et al., 2017, p. 2). Furthermore, the main regional intergovernmental organization, Arctic Council, is still lacking a strong position on renewable energy development. Though recent developments[8] suggest that this is starting to change, a more coherent positioning on the issue needs to occur to encourage renewable energy developments in the region in a more consistent way.

There are, however, examples suggesting that corporations are looking to use power procurement strategies to cut CO_2 emission in on-grid parts of the Arctic. The energy company, ENI, has connected their operations in the Goliat-field in the Barents Sea to the Nordic power grid through subsea cable, which is estimated to reduce 50 percent of the CO_2 emissions from the platform (Eni Norge, 2014). However, other offshore energy companies are not following suit as power cuts are making this solution less desirable (Nilsen, 2016). Statoil recently opted for a gas-fired power solution to their operation in the same area (Staalesen, 2017). This suggests that technological innovations at the micro-level needs further development to handle the harsh conditions in some parts of the region, which requires more investment.

Conclusion

In conclusion, drawing on existing knowledge of energy transition mechanisms, this chapter has demonstrated that climate change has influenced corporations' energy procurement preferences and patterns, which in turn is contributing to the renewable energy transition.

The findings suggest that there is a vast potential in procurement practices for corporations and governments wishing to enhance the transition toward a low-carbon future. While renewable energy development is challenging in the Arctic, this chapter has highlighted the need for corporate champions that will raise awareness, invest in infrastructure and demand access to renewable energy for their operations in this region. It also points to the need for a stronger political positioning and framework on the matter.

Further research and reflections are necessary to enhance the understanding of the renewable energy transition. This study has shed light on corporations' role in this process, which hopefully can contribute to further discussions, innovations and practical energy solutions.

Notes

1 The Rocky Mountain Institute's Business Renewable Center has tracked new corporate renewable energy deals made in the US and Mexico since 2012. See here: http://busi nessrenewables.org/corporate-transactions/.
2 Apple, Autodesk, Dentsu Aegis, Equinix, Facebook, Interface, Microsoft, TD Bank Group, Tetra Pak, Wells Fargo.
3 Referred to the environmental characteristics of a renewable energy resource, such as the emission data, represented in a corresponding certificate (Mass Energy, NA).
4 One unit consisting of the energy commodity and the green attributes (Mass Energy, NA).
5 World Business Council for Sustainable Development (WBCSD) and Renewable Energy Buyers Alliance (REBA) are similar business collaborations and ShareAction is an organization helping investors take into consideration social and environmental aspects in their investments.
6 Yeldcos are dividend-driven spin-offs of larger parent companies.

7 See the Pembina Institue report "The True Cost of Fuel in the Arctic: Power Purchase Policies, Diesel Subsidies and Renewable Energy" (Lovekin et al, 2016).
8 The Arctic Council Sustainable Development Working Group (SDWG) recently established the Arctic Remote Energy Network Academy and the Arctic Renewable Energy Atlas (SDGW, 2015), and region-specific feasibility studies have been carried out (Arctic Council, 2017).

References

Abbot, K. W. and Snidal, D. (2009) "Strengthening International Regulation Through Transnational New Governance: Overcoming the Orchestration Deficit" in *Vanderbilt Journal of Transnational Law* Vol. 42, No. 2, pp. 501–578.

Allen, R. (2012) "Backward into the Future: The Shift to Coal and Implications for the Next Energy Transition" in *Energy Policy* Vol. 50, pp. 17–23.

Arctic Council (2017) *Renewable Energy: Investments in the Arctic.* Available at: https://oaarchive.arctic-council.org/bitstream/handle/11374/2040/2017-06-13-ACAP-fact-sheet-renewable-energy-investments-in-the-arctic-letter-size-ENGLISH-DIGITAL.pdf?sequence=1&isAllowed=y (Accessed: January 12, 2018).

Beavers, R. (2006) "Corporate Solar is Coming" in *Google Blogpost,* 16 October 2006. Available at: https://googleblog.blogspot.no/2006/10/corporate-solar-is-coming.html. (Accessed: August 20, 2016).

Berdin, V. Kh., Kokorin, A. O., Yulkin, G. M. and Yulkin, M. A. (2017) *Renewable Energy in Off-grid Settlements in the Russian Arctic.* WWF Russia: Moscow, p. 2.

Bloomberg New Energy Finance (BNEF) (2017) *Clean Energy Investment End of Year 2016.* Available at: https://about.bnef.com/clean-energy-investment/#toc-download (Accessed: June 30, 2017).

Buckley, T. (2016) "In Emerging Economies, New Forms of Renewable-Energy Financing Are Taking Root" in *IEEFA Update,* December 19, 2016. Available at: http://ieefa.org/emerging-economies-new-forms-renewable-energy-financing-taking-root/ (Accessed: August 15, 2017).

Bumpus, A. G. (2015) "Firm Responses to a Carbon Price: Corporate Decision Making under British Columbia's Carbon Tax" in *Climate Policy* Vol. 15, No. 4, pp. 475–493.

Business Renewables Center (2018) "BRC Deal Tracker" by *The Rocky Mountain Institute.* Available at: http://businessrenewables.org/corporate-transactions/ (Accessed: January 14, 2018).

Campbell, J. (2007) "Why Would Corporations Behave in Socially Responsible Ways? An Institutional Theory of Corporate Social Responsibility" in *Academy of Management Review* Vol. 32, No. 3, pp. 946–967.

CDP (2017) *Climate Change 2016 Information Request Alphabet, Inc.* Available at: https://static.googleusercontent.com/media/www.google.com/en//green/pdf/climate-change-2016-information-request-alphabet-inc.pdf (Accessed: July 6, 2017).

Clean Energy Ministerial (CEM) (2016a) *Corporate Sourcing of Renewables.* Available at: www.cleanenergyministerial.org/Our-Work/CEM-Campaigns/Corporate-Sourcing-of-Renewables (Accessed: July 6, 2017).

Clean Energy Ministerial (CEM) (2016b) *Public-Private Roundtables at the Seventh Clean Energy Ministerial.* Available at: www.cleanenergyministerial.org/Portals/2/pdfs/CEM7_Roundtables_Report_Web_Version_Final.pdf (Accessed: July 6, 2017).

Clean Energy Ministerial (CEM) (2016c) "Corporate Sourcing for Renewables Campaign Launches at Seventh Clean Energy Ministerial" in *News by Clean Energy Ministerial,*

June 2, 2016. Available at: www.cleanenergyministerial.org/News/corporate-sourcing-for-renewables-campaign-launches-at-seventh-clean-energy-ministerial-68649 (Accessed: June 11, 2017).

Clean Energy Ministerial (CEM) (2017a) "About the Clean Energy Ministerial" on *Official Website*. Available at: www.cleanenergyministerial.org/About (Accessed: July 6, 2017).

Clean Energy Ministerial (CEM) (2017b) "Campaigns" on *Official Website*. Available at: www.cleanenergyministerial.org/Our-Work/Campaigns-Index (Accessed: July 6, 2017).

Correia, F., Howard, M., Hawkins, B., Pye, A. and Lamming, R. (2013) "Low Carbon Procurement: An Emerging Agenda" in *Journal of Purchasing & Supply Management* Vol. 19, No. 1, pp. 58–64.

Crespin-Mazet, F. and Dontenwill, E. (2012) "Sustainable Procurement: Building Legitimacy in the Supply Network" in *Journal of Purchasing & Supply Management* Vol. 18, No. 4, pp. 207–217.

Donnelly, G. (2017) "Google Just Bought Enough Wind Power to Offset 100% of Its Energy Use" in *Fortune*. Available at: http://fortune.com/2017/12/01/google-clean-energy/ (Accessed: June 6, 2017).

Duhaime, G. and Caron, A. (2017) "Chapter 2: The Economy of the Circumpolar Arctic" in S. Glomsrød, G. Duhaime and I. Aslaksen (Eds.) *The Economy of the North 2015.* Oslo, Norway: Statistics Norway, p. 17.

Dunlop, S. and Roesch, A. (2016) "Power Purchase Agreements (PPAs) Supply Contract Business Model" in *EU-Wide Solar PV Business Models: Guidelines for Implementation a SolarPower Europe's PV Financing Projects Report.* Solar Brussels, Belgium: Power Europe Brussels.

Eckhouse, B. (2017) "Corporations Key Demand Drivers for Wind, Solar, Moody's Says" in *Renewable Energy World*. Available at: www.renewableenergyworld.com/articles/2017/03/corporations-key-demand-drivers-for-wind-solar-moody-s-says.html (Accessed: November 17, 2017).

Eni Norge (2014) *Information About: Installation of a Submarine Cable to the Goliat Field in the Barents Sea.* Available at: https://issuu.com/newsonrequest/docs/eni_norway_-_installation_of_a_subm (Accessed: June 11, 2017).

Finnemore, M. (1996) *National Interests in International Society.* Cornell University Press: Ithaca and London, pp. 29, 30.

Foreign Policy (2016) *Diplomat of the Year Award 2016*, video recording, YouTube. Available at: www.youtube.com/watch?v=q0HUMejthUM (Accessed: November 17, 2016).

Fouquet, R. (2010) "The Slow Search for Solutions: Lessons from Historical Energy Transitions by Sector and Service" in *Energy Policy* Vol. 38, No. 11, pp. 6586–6596.

Geels, F. W. and Schot, J. (2007) "Typology of Sociotechnical Transition Pathways" in *Research Policy* Vol. 36, No. 3, pp. 399–417.

Global Wind Energy Council (N/A) "Wind in Numbers" on *Official Website*. Available at: www.gwec.net/global-figures/wind-in-numbers/ (Accessed: January 14, 2018).

Goldthau, A. and Sovacool, B. (2012) "The Uniqueness of the Energy Security, Justice, and Governance Problem" in *Energy Policy* Vol. 41, pp. 232–240.

Google (2013) *Google's Green PPAs: What, How, and Why.* Available at: https://static.googleusercontent.com/media/www.google.com/en/us/green/pdfs/renewable-energy.pdf (Accessed: January 14, 2018).

Google (2017) *Environment Blog.* Available at: https://blog.google/topics/environment/ (Accessed: December 10, 2017).

Grubler, A. (2012) "Energy Transitions Research: Insights and Cautionary Tales" in *Energy Policy* Vol. 50, pp. 8–16.

Hoejmose, S. U. and Adrien-Kirby, A. (2012) "Socially and Environmentally Responsible Procurement: A Literature Review and Future Research Agenda of a Managerial Issue in the 21st Century" in *Journal of Purchasing & Supply Management* Vol. 18, No. 4, pp. 232–242.

Hope, C., Whiteman, G. and Crawford-Brown, D. (2015) "3 Industries That Will be Hit by Arctic Change" in *World Economic Forum Agenda*. Available at: www.weforum.org/agenda/2015/09/3-industries-hit-by-arctic-change/ (Accessed: December 12, 2017).

International Energy Agency (IEA) (2016) *Medium-Term Renewable Energy Market Report 2016*. OECD: Paris.

International Energy Agency (IEA) (2017) *Electricity*. Available at: www.iea.org/topics/electricity/ (Accessed: December 12, 2017).

International Governmental Panel on Climate Change (IPCC) (2014) *Climate Change 2014 Synthesis Report Summary for Policymakers Chapter*. Available at: www.ipcc.ch/pdf/assessment-report/ar5/syr/AR5_SYR_FINAL_SPM.pdf (Accessed: December 12, 2017).

International Renewable Energy Agency (IRENA) (2016) *Innovation Outlook: Offshore Wind*. International Renewable Energy Agency: Abu Dhabi, pp. 1, 2, 4, 5.

International Renewable Energy Agency (IRENA) (2017) *REthinking Energy 2017*. International Renewable Energy Agency: Abu Dhabi, pp. 64, 68, 69, 70.

Kimball, A. (2016) "Innovating for a Cleaner Energy Future" in *Google Blogpost*. Available at: https://blog.google/topics/public-policy/innovating-for-cleaner-energy-future/ (Accessed: December 11, 2017).

Lovekin, D., Dronkers, B. and Thibault, B. (2016) *The True Cost of Fuel in the Arctic: Power Purchase Policies, Diesel Subsidies and Renewable Energy*. The Pembina Institute: Canada.

March, J. and Olsen, J. (1998) "The Institutional Dynamics of International Political Orders" in *International Organization* Vol. 52, No. 4, pp. 943–969.

Mass Energy (NA) *Glossary*. Available at: www.massenergy.org/renewableenergy/glossary#BundledRenewableElectricityProduct) (Accessed: September 11, 2017).

McMahon, J. (2016) "Renewable Energy is Now Inevitable, Energy Secretary Says, Citing Price" in *Forbes*. Available at: www.forbes.com/sites/jeffmcmahon/2016/06/03/renewable-energy-inevitable-energy-secretary-says-because-of-plunging-prices/#11286d8f3a8b (Accessed: December 12, 2017).

Moodie, A. (2016) "Google, Apple, Facebook Race towards 100% Renewable Energy Target" in *Guardian*, December 6, 2016. Available at: www.theguardian.com/sustainable-business/2016/dec/06/google-renewable-energy-target-solar-wind-power (Accessed: December 12, 2017).

Nilsen, T. (2016) "Goliat Remains Shut Down, Norway Orders Plan on How to Avoid Power Outage" in *The Independent Barents Observer*.

Nordregio (2011) *Generation of Electricity in the Arctic*. Available at: www.nordregio.se/en/Maps/05-Environment-and-energy/Generation-of-electricity-in-the-Arctic/ (Accessed: August 20, 2017).

Norwea (2016) *Google and BlackRock Invest in Norwegian Wind Farm Tellenes*. Available at: www.norwea.no/nyhetsarkiv/visning-nyheter/google-and-blackrock-invest-in-norwegian-wind-farm-tellenes.aspx?PID=1145&Action=1 (Accessed: June 10, 2017).

Norwegian government (2017) *Norway's Arctic Strategy – between Geopolitics and*

Social Development. Norwegian Ministry of Foreign Affairs and Norwegian Ministry of Local Government and Modernisation: Oslo, p. 32.

Pintea, M.-O., Stanca, L., Achim, S.-A. and Pop, I. (2014) "Is There a Connection Among Environmental and Financial Performance of a Company in Developing Countries? Evidence from Romania" in *Procedia Economics and Finance* Vol. 15 pp. 822–829.

Renewable Energy 100 (RE100) (2017) *Annual Report: ACCELERATING CHANGE: How Corporate Users are Transforming the Renewable Energy Market.* Climate Group and CDP, pp. 1, 3, 5, 20, 21.

Rip, A. and Kemp, R. (1998) "Chapter 6: Technological Change" in S. Rayner and E. L. Malone (Eds.), *Human Choice and Climate Change. Vol. II, Resources and technology.* Battelle Press: Columbus, OH, p. 338.

Rotmans, J., Kemp, R. and van Asselt, M. (2001) "More Evolution than Revolution: Transition Management in Public Policy" in *Foresight* Vol. 3, No. 1, pp. 15–31.

Royal, H. (2016) "What's the Difference between a Traditional and Renewable PPA" in *renewable choice ENERGY.* Available at: www.renewablechoice.com/blog-difference-between-traditional-and-renewable-ppa/ (Accessed: November 3, 2017).

Rugman, A. and Verbeke, A. (1998) "Corporate Strategies and Environmental Regulations: An Organising Framework" in *Strategic Management Journal* Vol. 19, pp. 363–375.

Shove, E. and Walker, G. (2007) "CAUTION! Transitions Ahead: Politics, Practice, and Sustainable Transition Management" in *Environment and Planning* Vol. 39, No. 4, pp. 763–770.

Smil, V. (2016) "Examining Energy Transitions: A Dozen Insights Based on Performance" in *Energy Research & Social Science* Vol. 22, pp. 194–197.

Solar Power Authority (2016) *A Guide to Solar Roof Tiles: The Next Big Thing.* Available at: www.solarpowerauthority.com/guide-to-solar-roof-tiles/ (Accessed: December 11, 2017).

Sotos, M. (2014) "Scope 2 Guidance Case Studies: Organizations Creating, and Applying the Results of, GHG Inventories Based on the GHG Protocol Scope 2 Guidance" by *Greenhouse Gas Protocol*, p. 12.

Sovacool, B. (2016) "How Long Will it Take? Conceptualizing the Temporal Dynamics of Energy Transitions" in *Energy Research & Social Sciences* Vol. 13, pp. 202–215.

Staalesen, A. (2017) "Statoil Announces Massive Investment in its Northernmost Oil Field" in *Arctic Now*, December 5, 2017. Available at: www.arcticnow.com/business/energy/2017/12/05/statoil-announces-massive-investment-in-its-northernmost-oil-field/ (Accessed: December 12, 2017).

Sustainable Development Working Group (SDGW) (2015) *PROJECTS 2015–2017.* Available at: www.sdwg.org/activities/current-projects/ (Accessed: December 15, 2017).

The Energy Gang (2016) *The Art Of The Deal: How Corporates Are Investing In Renewables*, November 17, 2016. Available at: https://soundcloud.com/the-interchange/art-of-the-deal-why-fortune-100-companies-are-buying-so-much-renewable-energy (Accessed: May 20, 2017).

United Nations (UN) (2015) *Paris Agreement.* Available at: http://unfccc.int/files/essential_background/convention/application/pdf/english_paris_agreement.pdf (Accessed: May 2, 2017).

Yergin, D. (2013) "Executive Perspective: Daniel Yergin on the Puzzle of Energy Transitions" in *Thomson Reuter's Sustainability*, March 13, 2013. Available at: http://sustainability.thomsonreuters.com/2013/03/13/executive-perspective-daniel-yergin-on-the-puzzle-of-energy-transitions/ (Accessed: July 16, 2017).

Part V

Arctic energy policies and standards

9 Achieving high performance with business process and quality management within the Arctic energy industry

Melania Milecka-Forrest

Business process management and its benefits for Arctic energy companies

Business process management (BPM) is a field of operations management that focuses on improving corporate performance by managing companies' business processes. It is an integrated part of management and it is a programme that should be maintained on continuous basis. Many current users of BPM recognise technology as its key component. However, if you take a simple view on the concept of BPM it becomes apparent that management of processes can improve technical as well as managerial issues within businesses.

BPM can be supported by software. In order to successfully adopt BPM it is necessary to conduct regular end-to-end review of organisational processes, before and after the implementation. A well-designed and managed BPM can be recognised as a valuable asset that through the improvement and reengineering of processes can fuel the delivery of corporate objectives.

BPM aims to improve operational business processes with or without technologies. By modelling processes and analysing those using scenario based case studies, simulations and gamefication approaches, management is able to gather all necessary information on how to reduce costs while improving level of engagement with key company stakeholders. BPM is also often associated with technology that helps manage and control operational processes. BPM is therefore a collaborative effort between business units and technology that leads into creation of efficient, effective and logical processes. BPM gives you greater control over operational processes and makes your organisation more responsive to continuously changing environment.

Key benefits of adopting BPM by Arctic energy companies

BPM helps leaders and organisations improve performance thorough the variety of approaches such as analysis, careful design, vigilant observation, control and remodelling of processes. Adoption of BPM by Arctic energy companies provides organisations with wide range of long and short term benefits.

Arctic business environment is in a state of change. According to the Arctic Human Development Report (2004), Arctic societies nowadays face an

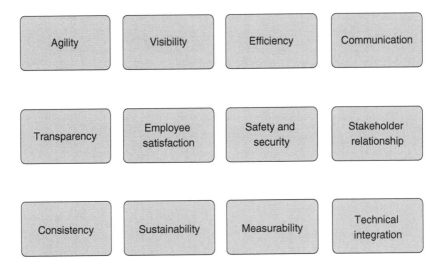

Figure 9.1 Benefits of adapting BPM.

unprecedented combination of rapid and stressful changes involving environmental processes (e.g. the impacts of climate change), cultural developments
(e.g. the erosion of indigenous languages), economic changes (e.g. the emer-
gence of narrowly based mixed economies), industrial developments (e.g. the
growing role of multinational corporations engaged in the extraction of natural
resources) and political changes (e.g. the devolution of political authority).

Recent discussions with regards to current and future developments in the
Arctic highlight the importance of local stakeholders including local people,
communities and economies in business decision making. It has become increasingly important that business processes adapt to new conditions, approaches and
requests influenced by a wide range of stakeholders. BPM helps organisations
become more agile, adapt to changes and deploy the best option that while sta-
bilising the system integrates new systems. BPM facilitates the design of processes that are flexible. It can also facilitate automation of repetitive activities
that not only boosts visibility but also allows organisations to increase productivity by reducing areas of waste. Automation will help achieve higher output
since employees focus on spending more time on more important core functions.

Increased visibility within an organisation will allow leaders to concentrate on
inefficiencies. BPM allows organisations to work more efficiently while reducing
costs, saving their resources and, most importantly, reducing risks. In order to
maximise efficiency, an integration of organisation's processes from start to finish
is enormously important. Moreover, implementing the right BPM strategy can
significantly increase knowledge exchange within an organisation. While companies can carefully monitor their processes and look for ways of maximising
benefits, BPM can allow the development of communication channels stimulating

the collaboration between private and public sectors, community and academic research expert teams that can help explore potential solutions based on energy issues and adopt the most beneficial to stakeholders.

In recent years it became evident that organisations need to be more transparent with regards to compliance with standards, regulations and policies. BPM allows organisations to comply with regulatory requirements and make swift changes to avoid any penalties and fines. Compliance can be integrated into the life cycle process that also enhances transparency of all stages of processes to external and internal bodies.

While a wide range of innovative energy production, storage and transfer technology systems is under development, new technology requires significant adaptation and improvements to current infrastructure. BPM allows agility in responding to change adaptation. Any changes to processes would, in effect, include modification of compliance, safety and security activities. Any key stakeholders affected would therefore need to be engaged in specific training and development activities that can also be an integral part of BPM. People need to have specialist training to get the more attractive jobs in the energy sector. Without the adequate training, job opportunities offered in the energy sector will mostly benefit educated people from other regions than the Arctic (Mikkelsen and Langhelle, 2008).

Mikkelsen and Langhelle (2008) highlight the need to train local people, not just for employment purposes but also to ensure that relevant people are aware of the consequences of not following specific safety standards. Employment and training of local people not only increases their standard of living but also increases their awareness of the dynamics of the energy industry. On the other hand, since automation takes away all repetitive tasks, employees can focus on their more engaging work activities. With transparent processes, continuous monitoring and evaluation tracking, and engagement with key stakeholders, especially local communities, information becomes easily accessible, which in turn makes employees an active part of increased productivity and quality standards.

Energy plans within organisations should have a clear vision and strategy in order to encourage collaboration and participation of key stakeholders. Organisations' support in the development of community networks and collaborative initiatives with national and international partners can enhance organisations' reputation and longevity. Early engagement of key stakeholders in the decision making processes and throughout all levels of planning may allow organisations to futureproof their actions. Collaboration with all relevant stakeholders including local communities can help eliminate risks and produce mutual benefits including more support in future developments in the area. BPM's flexibility allows leaders to incorporate such important relationships within business processes.

The key principle of BPM is to allow continuous improvement of processes. Organisations working in a changing environment with changing conditions are able to adopt new practices swiftly in order to deliver sustainable results. The implementation of new activities within processes can be successfully achieved with BPM while still maintaining standards and meeting key performance indicators (KPIs) within systems and ensure these activities are executed the way they were planned and designed. Processes that are standardised and employees with clearly defined roles are accountable for bringing less variation and providing effective consistency.

BPM allows all processes to be measured from end-to-end results and can be compared to past practices. This doesn't only help manage business productivity but also helps manage employee and process performance as well as measure the impact of highs and lows on key stakeholders and shareholders. With the use of appropriate software, the reporting of performance to leaders allows them to make suitable long and short term decisions.

BPM systems also allow leaders to receive key information swiftly and in the format needed. It allows the implementation of frequent process changes while maintaining control. Team's and stakeholder's collaboration and reaction time can also be improved with use of data and information dashboards presented in user-friendly formats.

According to John Jeston (2018) the foundations of the BPM programme provides a solid operational focus visible across the organisation. Senior management is responsible for determining the organisation's strategy and ensuring that the business process contributes to its fulfilment. Processes that are aligned with the strategy are most effective in achieving business objectives and are more sustainable in the medium to long term. Clear business strategy is required to sustainably maximise the benefits of Arctic business dynamics.

Business process management strategy

BPM strategy within the Arctic environment depends entirely on the concept of process management chosen and implemented within an organisation. There are multiple ways of approaching BPM strategy. The more simplified approach would require consideration of a series of steps:

a   understand the key purpose for the implementation of BPM;
b   model and formalise fundamental business processes;
c   understand the value chain, its key components and factors affecting it;
d   review key technological resources available and select the most appropriate ones;
e   conduct benefits vs. risks impact analysis on local communities;
f   conduct external and internal impact analysis to find strengths, weaknesses, opportunities and threats in organisations using reliable, currently available information;
g   redesign and finalise business processes;
h   cooperate with multiple stakeholders on the development of a set of integrated, science-informed, ecosystem-based, flexible regulations;
i   proactively manage, monitor and control process activities, and where necessary adopt new solutions;

j   simulate and test potential improvements in a safe and controlled environment;

k   communicate top management decisions down the organisation's hierarchy structure using clear and transparent information backed with data and comprehensive reports;

l   review business processes on a regular basis using reliable and up to date information.

The more complex approach would recommend that managers refer to Burlton's Hexagon when developing their strategies. As Roger Burlton (2010) specifies in his handbook on BPM:

> The Burlton Hexagon shows that processes are the mechanisms that are measurable and deliver performance through the definition of the process KPIs in support of the stakeholder relationship and corporate objectives. It also shows that work flows by themselves are not sufficient.
>
> (Burlton, 2010)

The processes must also consider the constraints or empowerment delivered by policies and rules, software technologies, facilities, all aspects of human capital, human motivation and organisation design. In his argument, Dunning (2003) supports the importance of trust, reciprocity and due diligence within organisations operating in changing world:

> We are moving out of an age of hierarchical capitalism and into an age of alliance capitalism. This is placing a premium on the virtues needed for fruitful and sustainable coalitions and partnerships (be they within or among institutions), such as trust, reciprocity, and due diligence.
>
> (Dunning, 2003)

In order to achieve maximum benefits of this methodology, all components must be aligned and maintained. Moreover, the transparency of the alignment and management of components is crucial. This particular methodology benefits from inclusion of knowledge and information that is still in development and sometimes neglected in the Arctic energy sector.

Key challenges in defining a future proof strategy in the Arctic energy sector

There are a great variety of theoretical methodologies that can be tailored and adapted by organisations around the world. Arctic energy companies are facing multiple challenges, some of which are discussed at the Arctic Energy Summit every year.

The institutional landscape is changing in the Arctic. As argued by Young (2005), the Arctic has experienced a dramatic shift from the status of sensitive theatre of operations for the deployment of strategic weapons systems to that of focal point for a range of initiatives involving transnational cooperation. These initiatives take a variety of forms and involve different actors. Moreover, as Young (2005) observed, the establishment of the Arctic Environmental Protection Strategy in 1991 and the creation of the Arctic Council in 1996 involve international and diplomatic agreements. Other important initiatives in the social and scientific arena were the foundation of the Inuit Circumpolar Conference launched in 1977, the International Arctic Science Committee developed in 1990 and the University of the Arctic, launched in 2001, more focused on non-governmental activities and on addressing social and scientific Arctic challenges.

The important factor, however, is that Arctic cooperative arrangements have become a means for articulating regional interests and for protecting regional actors from the side effects of global business processes. In most cases, these efforts deal with the impacts of large-scale environmental changes and with the effects of economic and social globalisation (Young, 2005).

Following recent discussions and observations there are some common challenges and suggestions linked to approaches taken towards energy. The energy developments in the Arctic should balance the application of energy resources with the needs of northern communities in a way that the energy resources could be used for the benefit of all northern residents. As defined in the Arctic Energy Summit of 2015, the residents of the Arctic pay some of the highest energy prices in the world.

Any future initiatives in energy development should be consulted with a large amount of stakeholders to ensure that the most appropriate measures and approaches are taken to provide sustainable energy balanced with environmental protection and respect for cultures and traditional ways of living. Moreover, business organisations and government should invest in increasing access to energy training in order to address local employment needs as well as the equipment maintenance needs. It is essential to set up a transparent business system that will allow investment decisions to be made in line with public interest.

More emphasis should be on the knowledge of management concepts and the knowledge of management life cycle in support to organisations' objectives and BPM. This would help develop management foundations for organisations' capability to learn from information available inside and outside of the organisations.

According to Young (2005), the cooperative arrangements remain fragile in the sense that they have not yet 'hardened into a well-established assemblage of social practices featuring a common discourse that can turn expediential and transient calculations of political interest into a mode of operation that becomes a matter of second nature to its participants' (Young, 2005). In order to make it even more complicated, global environmental change and globalisation 'threaten to overwhelm efforts to carve out coherent agendas at the regional level and to pursue them without undue concern for the linkages between regional activities and planetary processes' (Young, 2005). It means that finding the right course of action and finding answers to the business and social challenges mentioned above became even more difficult.

Energy business process management life-cycle

The key value of BPM, in this context, is the creation of collaboration and awareness within and outside of organisations. The main initiative of the BPM system is to look for opportunities to improve business processes while encouraging and strengthening the interaction with stakeholders.

The most common representation of BPM life-cycle includes five phases: design, modelling, execution, monitoring and optimising.

Process design first covers areas such as identification of existing processes, modelling and formalisation of new processes. The focal point would be the process flow, activities involved and operating procedures.

Process modelling takes into consideration all the proposed tasks, activities and flows from the design stage and uses simulation to stimulate different conditions triggered by external and internal business environment changes.

At the execution stage, the development of proposed business processes takes place. It involves the combination of software and human activities being either run separately or combined. It is a particularly challenging stage and thorough

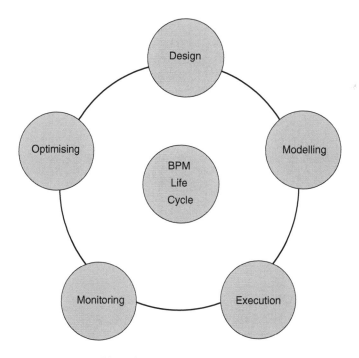

Figure 9.2 BPM life cycle.

research into the software and technology available is recommended in order to create the most effective and user-friendly BPM system possible.

The process monitoring phase encompasses performance measures that will map out all the activities that took place within the process. This stage is particularly beneficial for leaders. It will allow them to gather all necessary information needed to trigger improvements within an organisation. Full comprehensive reports can be created and used for negotiations with external stakeholders.

At the process optimising stage we should be able to extract all the data and information known about the current end-to-end business processes. It includes potential issues such as bottlenecks, potential changes to the business environment that impact the cost/profits, availability of resources and stability of the business.

The BPM life-cycle is very simple and allows a lot of room for further development, interpretation and flexible adaptation. This gives business managers freedom to review their strategy, and to design their processes at all stages of the cycle in the way that is most appropriate for their particular business case. The lack of sufficient information on the life-cycle can also leave business managers confused about how to design a process that will meet all their strategic objectives and still be positively perceived by their teams and stakeholders. Another pitfall of this cycle is that it does not include any space for knowledge collection and research activities if it is a purely hard version of step-by-step process design life-cycle.

In an ideal scenario, business strategy should cascade down and feed into BPM activities. We are therefore able to expand the BPM life cycle and include key strategic considerations for the specific Arctic context.

The design stage would therefore also include identification of business drivers, objectives, purpose and strategic intent. Stakeholders' identification, liaison and management would also be recommended. Collaboration on the development of a set of integrated, science-informed, ecosystem-based and flexible regulations, as well as past performance results analysis and the consideration of lessons learnt log, would be encouraged.

The modelling stage would include the consideration of the effect of processes on the value chain, conducting benefit and risk impact analysis on local communities alongside external and internal impact analysis. The modelling would help recognise key strengths, weaknesses, opportunities and threats in organisations by using reliable, currently available information that would also help develop case scenarios and simulations.

The execution stage would include design of roles for software and people, review of key technological resources available and selection of the most appropriate ones. This stage would allow the redesign and completion of business processes and, finally, top management would be able to communicate their decisions down the organisations' hierarchy structure using clear and transparent information backed with data and comprehensive reports.

The monitoring stage would include a proactive approach to management, monitoring and control of process activities. Adoption of new solutions, simulations and testing of potential improvements in a safe and controlled environment would also be recommended.

The optimising stage would include some of the similar activities as the ones within the execution and monitoring stages, for example, simulation and testing potential improvements in a safe and controlled environment, using more in-depth information. It would also comprise the review of business processes using reliable and up-to-date information and sharing information with stakeholders to seek help in designing new solutions. Top management should be able to communicate key decisions down the organisations' hierarchy structure using clear and transparent information backed with data extracted from lessons learnt to inform future decision-making.

Throughout all stages of the cycle, knowledge creation and sharing opportunities should be made available to all stakeholders, increasing process transparency.

A complex set of factors influence the interaction and knowledge sharing within a group of people with the forces of the global economy and the outcomes they experience. This has a particular impact on the Arctic business environment. These factors include (Mikkelsen and Langhelle, 2008):

a   The impact of the 'state' at all levels and the 'civil sector' on the multiple overlapping modes of social regulation and, therefore, on participants in the global economy, and the influence of these participants on the 'state' and the 'civil sector'.
b   The community-in-question's approach to economic development including its history, current circumstances, objectives and approach to participation in the global economy, including strategies for participation, transformation and exclusion and actual outcomes.
c   Corporate (as the usual representative of the regime of accumulation encountered by communities) responses to the community-in-question. Particularly motivating forces include the community's control over the critical natural, human and financial resources. Community members' attractiveness as a market and corporate strategies and objectives may also affect outcomes. Outcomes, of course, influence the ongoing process of development.
d   The expected mode of development and the actual mode that emerges in a particular circumstance.

In order to allow knowledge creation within and outside of an organisation, business leaders should promote knowledge exchange in a variety of ways that are well perceived by employees, communities and all the relevant stakeholders.

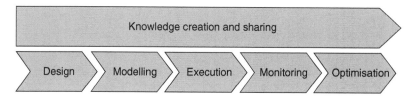

Figure 9.3 Integration of knowledge creation and sharing with BPM life cycle.

Business leaders should encourage knowledge sharing opportunities by not only preventing any knowledge loss but also by increasing the effectiveness of formal and informal knowledge acquisition and, as result, enable skills development and process improvement on continuous basis.

As far as the internal advantages for the enterprise are concerned, free flow of information inside the organisation would enhance exchange of experiences among employees and project teams (Bitkowska, 2016) enhancing the development of employees' skills, based on their experiences and motivation. Benefits from investing in employees' creativity and ideas generate innovative solutions. Changes in terms of the flow of information between departments in a company, resulting from the introduction of appropriate procedures, stimulate positive reaction to changes occurring in the environment providing faster decision-making.

External benefits of knowledge management encompass the fundamental development of an organisation, the creation of product improvements, the increase in organisations' competitiveness and continuous improvement of employees' qualifications. In this path, companies would be able to respond quicker to market needs by allowing continuous observation of clients' needs and the activities of the key competitors. Investment in knowledge exchange and management would introduce a culture of continuous enhancement of the quality of products or services and the creation of new ones. Finally, significant changes in the shaping of the relationships with clients, suppliers and partners in the market would also significantly enhance organisations' image and reputation.

Development of business strategy that includes serious considerations of knowledge management and alignment of business processes with organisational knowledge management strategies will have considerable benefits for Arctic organisations. It will help those organisations overcome a diversity of issues currently raised by international bodies and Arctic Energy Summits.

The role of quality management within energy sector

Peter Senge *et al.* (2007), in his *The Fifth Discipline Fieldbook*, defines quality as a transformation in the way we think and work together, in what we value and reward, and in the way we measure success. All stakeholders collaborate to design and operate a seamless value adding system that incorporates quality control, customer service, process improvement, supplier relationships and good relations with the communities.

Many quality experts have contributed to the creation of quality management theories and practices and, therefore, due to the great variety of tools and methods available, many organisations end up choosing and selecting tools and methods that correspond adequately to their organisations' needs and approaches.

Many authors, including Peter Senge *et al.* (2007), recognise quality management as a system that looks mainly at factors contributing to the improvement of general business processes. Within an Arctic business environment there are additional and crucial areas that also have to be considered when reviewing quality and the impact of internal and external processes. There are key questions to be asked: for example, how to achieve sustainable development of energy supply that is reliable, economical and environmentally compatible to changing local contexts and how to prevent adverse impacts to environmental and society.

It is crucial to define customised solutions that provide, at all stages, environmental protection and health and safety standards, and that define how to conserve natural resources with respect to communities' cultures and traditions. There are still answers to be found on how to best employ knowledge and experience of global experts in the field to improve quality and environmental protection through the development of a comprehensive management system and relevant business processes. The extent of organisations' understanding of cultural and social vulnerabilities and resilience is also not clear.

Quality management is mainly related to managers' ability to develop strategies leading into building a responsive organisation. This can be achieved by adopting a continuous improvement mindset at every stage of managing and developing business processes. The concept of continuous quality management can very easily turn into a robotic set of activities as it focuses on standardising activities to ensure consistency. Within the Arctic energy environment it is crucial to maintain certain levels of creativity and innovation in quality management as this particular area is not yet fully developed. It is important that organisations focus on long term benefits by supporting energy supply for the future generation in a manner that does not cause negative impacts on the local environment. Preserving local natural resources will provide energy organisations with long term gains.

Strategic approach to quality management and continuous improvement can bring many benefits to organisations:

a   increase of productivity;
b   lower costs and increase of funds for future investment and development;
c   improved quality, therefore improved reliability with reduced waste;
d   improved employee satisfaction;
e   improved relationships with communities;
f   reduced pollution and health and safety risks;
g   clear processes and a set of regulations, procedures and systems;
h   assurance of best practices for future developments;
i   identification of key trends and resolving key problems.

The key steps in implementation of quality management and continuous improvement approaches consist of:

a   Top management ability and commitment to quality and continuous improvement. Inclusion and identification of quality management and continuous improvement as one of organisation's strategies.

b   Business's transparent and clear intent, motivation, set of core values and principles.
c   Developing a comprehensive plan broken down into key indicators for each level within organisational structures. The plan should clearly identify market, customer demand and stakeholders needs.
d   Mapping out and modelling of processes leading into quality management and continuous improvement activities.
e   Keeping all stakeholders in the loop and engaging them where necessary for maximum benefits. This would include investment in internal staff development activities.
f   Monitoring business process performance on a continuous basis using transparent and unambiguous set of metrics.
g   Developing a continuous open feedback, awareness and open communication culture across all levels cascading down the organisations' structures. Evaluating and reviewing business processes a on regular basis.

A great example for cascading down quality management and continuous improvement activities is the statement and promises defined at corporate level at Siemens (2009b) within their Energy Sector Management Policy:

Responsibility for quality, as well as for environmental, health, and safety issues at Siemens is defined at corporate level. The CEO of the Energy Sector assumes responsibility for the stipulation and pursuance of the sector objectives relating to quality, environmental protection, health and occupational safety. The sector CEO ensures that the management system is developed and put into practice, and that its efficiency is continually improved. This includes:

a)   A clear direction for management and staff, and the motivation to work consistently in a Customer oriented manner. Compliance with customer requirements and the statutory and regulatory requirements are in the foreground.
b)   Definition of the Management Policy with the objective of increasing both customer benefits and economic value added.
c)   Regular management reviews to assess the effectiveness of the management system and initiation of measures to boost efficiency, thereby also improving product quality.
d)   Securing the availability of suitable resources.

The heads of the organizational units are responsible for all activities associated with quality, environmental protection, health, and occupational safety. They define the objectives for quality, environmental protection, and health and safety, more specifically and in greater detail, and define the respective areas of responsibility and authority. Apart from this, they also assume responsibility for the quality of their processes and products to the same extent as for compliance with the requirements for environmental protection and occupational health and safety. They decide on measures that can improve the quality of a product, and reduce a product's environmental impact, and ensure occupational health and safety at the workplace.

(Siemens, 2009b)

Siemens' policy is a great example of trust between all members of the organisation with a clearly defined set of accountabilities. All managers align their activities to the corporate goals and objectives.

Integration and alignment of quality management with business processes

As argued by Glomsrød and Aslaksen (2006) in *The Economy of the North*, the Arctic currently supplies about 16 per cent of all oil and gas to the global economy and it has substantial reserves to keep up this pace for long time. Hence, parts of the Arctic are seriously committed to decarbonisation as well as tackling the severe impacts of climate change. Moreover, there is a real concern related to the substantial exploration of mineral reserves, indirectly connected to the large-scale emission of greenhouse gases as they are processed in coal-based and polluting smelters around the world, including some Arctic regions. Thus, the Arctic is not only affected by global-induced climate change but also the increasing industrial activities in the region tend to contribute their share to global warming – the reason why management is a crucial factor to be analysed in this context.

Moreover, eco-systemic changes and the reduction in weather predictability present serious challenges to human health and food security, representing a real threat to the survival of local cultures (ACIA, 2004, p. 11). These impacts take place in the context of other stressors, among them chemical pollution, land-use changes, habitat fragmentation, and cultural and economic changes – stressors invariably linked to energy production (Hansen *et al.*, 2008; Mikkelsen and Langhelle, 2008).

In order to minimise the negative impact of business processes in the Arctic, it is critical that business processes and management approaches are aligned with quality management and continuous improvement plans (Brocke and Rosemann, 2010). Every step of the BPM life cycle should be aligned with a clear set of quality measures and activities.

One way of doing this is to apply the 'DMAIC improvement cycle and framework' used for a variety of improvement initiatives within organisations. DMAIC is an abbreviation for 'Define, Measure, Analyse, Improve and Control' and it was inspired by earlier improvement method 'Plan, Do, Check, Act' (PDCA) also known as Shewhart Cycle (Shewhart, 1939). While PDCA is a very theoretical framework, DMAIC is a very practical one. In order to improve and control a process efficiently, all five phases need to be followed.

The first stage is responsible for the definition of goals for improvement and design activity, purpose, creating a team and acquiring funding. The 'Define' phase can benefit from tools shown in Figure 9.4.

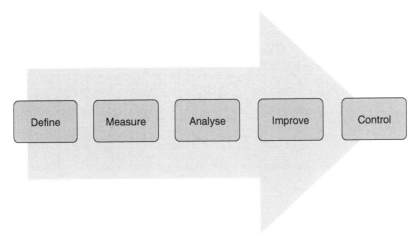

Figure 9.4 DMAIC.

The second stage is to measure the existing system. Within this stage it is important to develop a valid set of metrics to help monitor processes leading towards the achievement of goals identified in the 'Define' step. The 'Measure' phase can benefit from tools shown in Figure 9.5.

The third stage would involve analysis of the system to identify the gaps between desired and improved processes and current processes. The 'Analyse' phase can benefit from the tools shown in Figure 9.6.

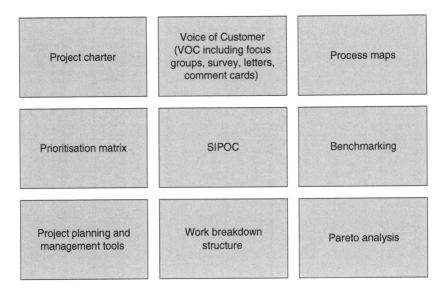

Figure 9.5 D for define.

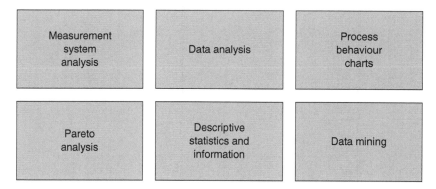

Figure 9.6 M for measure.

The fourth stage concentrates on the improvement of the processes, and finds new ways to better deliver processes with consistent completion of activities by using project management and other planning tools available to implement changes. The 'Improve' phase can benefit from the tools shown in Figure 9.7.

The final stage is to control newly designed business processes, by formalising them and by adopting appropriate management approaches towards staff and stakeholders. It also consists of aligning policies, procedures, guidance notes and a lessons learnt log. The 'Control' phase can benefit from tools shown in Figure 9.8.

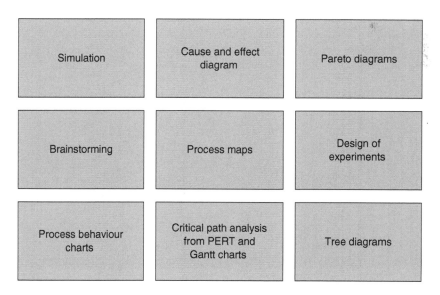

Figure 9.7 A for analyse.

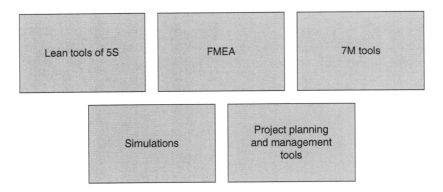

Figure 9.8 I for improve.

All of the above steps are crucial for a thorough review of processes within an organisation. This particular framework can be well used in the energy sector for improved outcomes. There are many available tools specified above that would help energy managers with their analysis and decision making responsibilities. To make the most of BPM and quality management strategies, business leaders should run or combine the explained frameworks.

All DMAIC (Define, Measure, Analyse, Improve and Control) steps combine well with BPA (Design, Modelling, Execution, Monitoring and Optimising) steps of the life cycle alongside all principles of knowledge management. Therefore, the framework developed earlier can be enhanced. The alignment and interaction between all three concepts will allow business leaders to make effective and carefully made decisions that not only consider hard business processes but also take into consideration the voice of the range of stakeholders encouraging collaborative efforts tending to result in more positive future outcomes.

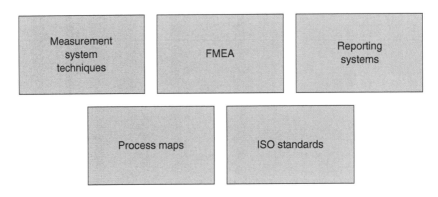

Figure 9.9 C for control.

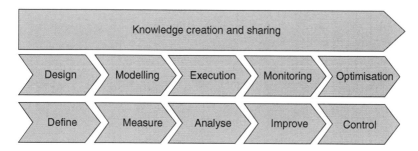

Figure 9.10 Integration of knowledge creation, BPM life cycle and DMAIC.

Conclusion

The Arctic is a very fragile area where small actions make a large-scale impact. It is critical that business leaders take very careful steps towards business process and quality management approaches. Every decision has to be carefully crafted with the support of all relevant stakeholders. Every decision needs to be based on the critical analysis conducted by using specific process and a quality management set of steps and tools. This chapter provided managers with key insights into BPM and also quality management theories. The key outcome of the discussed points is that every theory or framework can be adapted and customised according to the business needs. Business leaders have full flexibility in ensuring their decisions have minimal destructive impact and maximum benefits on business, communities and environment.

References

Arctic Climate Impact Assessment (ACIA) (2014) Accessed in January 2018 at: www. amap.no/arctic-climate-impact-assessment-acia.

Arctic Council (2004) *Arctic Human Development Report*, Akureyri, Iceland: Stefansson Arctic Institute.

Arctic Human Development Report (AHDR) (2004) Akureyri: Stefansson Arctic Institute.

Bitkowska, A. (2016) 'Knowledge Management vs Business Process Management in Contemporary Enterprises' *International Society for Manufacturing, Service and Management Engineering*, vol. 8, issue 2, pp. 31–37.

Brocke, J. and Rosemann, M. (2010) *Handbook on Business Process Management 2: Strategic Alignment, Governance, People and Culture*, Berlin, Germany: Springer.

Burlton, R. (2010) *Delivering Business Strategy through Process Management*, Berlin, Germany: Springer.

Dunning, J. H. (2003) *Making Globalization Good: The Moral Challenges of Global Capitalism*, Oxford: Oxford University Press.

Glomsrød, S. and Aslaksen, L. (2006) *Economy of the North, Project Report*, Oslo, Norway, Statistics Norway.

Hansen, O., Langhelle, O. and Anderson, R. (2008) *Arctic Oil and Gas: Sustainability at Risk?* Oxford, UK: Routledge.

Jeston, J. (2018) *Business Process Management: Practical Guidelines to Successful Implementations' Implementations*, Oxford, UK: Routledge.

Mikkelsen, A. and Langhelle, O. (2008) *Arctic Oil and Gas: Sustainability at Risk?* Oxford, UK: Routledge.

Senge, P., Kleiner, A. and Roberts, C. (2007) *Strategies for Building a Learning Organization*, New York: Crown Business.

Shewhart, W. A. (1939) *Statistical Method from the Viewpoint of Quality Control*, Dover, UK: Department of Agriculture.

Siemens (2009a) 'Quality Manual for The Energy Sector' Accessed in January 2018 on: w5.siemens.com/italy/web/pw/Chisiamo/Documents/SPA-Qualita/SPA/MQ_Energy_ ed5%20EN.pdf.

Siemens. (2009b) 'The Energy Sector Management Policy'. Accessed in January 2018 at: www.siemens.com/about/pool/en/businesses/energy/energy-sector-management-policy-en.pdf.

Young, O. R. (2005) 'Global Insights – Governing the Arctic: From Cold War Theatre to Mosaic of Cooperation', *Global Governance: a Review of Multilateralism and International Organizations*, vol. 11, issue 1, pp. 9–15.

10 Enhancing risk and safety management in large scale renewable energy generation

Natalia Rocha-Lawton

Introduction

The aim of this chapter is to explore the Arctic from a business perspective, focusing on risk and safety management in the context of renewable energy generation.[1] The change from fossil fuels to renewable energy has implications on how risk, safety and human capital is managed. The transformation of the industrial activity has repercussions on employment and how they are managed to cover the business strategy.

The Arctic renewable energy generation: risk environment and context

Economy in the Arctic

The Arctic has been characterised as a complex economic zone with valuable natural sources, social and physical environment in a state of constant change (Glomsrød and Aslaksen, 2009). This contrasting area in the world has a demographic mixture of remote inaccessible settlements and highly populated urban areas. The Arctic Council was established in 1996 after the Ottawa Declaration to provide the platform to promote cooperation among the eight Arctic States throughout their 28 regions with special emphasis on the environmental protection, sustainable development and particular Arctic issues (Arctic Council, 2017). The countries that participate as members in the Arctic Council are: Russian Federation, Sweden, Denmark (Greenland), Finland, Iceland, Norway, Canada and United States.

The region has a broad variety of business and industrial activities where traditionally fishing, mining, seafood, gas and oil are the most relevant activities (Loe *et al.*, 2014). In this context, the Arctic has attracted great interest for their vast natural sources to be exploited. The combined effects of global resource diminution, climate change and technological progress mean that the natural resource base of the Arctic – fisheries, minerals, and oil and gas – is now increasingly significant and commercially viable (Emmerson and Lahn, 2012) where the exploitation of oil and gas resources onshore and offshore in deeper and shallow waters are a growing activity in the Arctic region. Also, tourism is becoming one of the fastest growing sectors that will create new jobs for local people and expatriates.

The global energy demand is estimated to grow from 524 quadrillion BTU (British thermal unit) in 2010 to 630 quadrillion in 2020. New sources on energy supply will be needed to cover gas and oil areas in decline. Along these lines, Loe *et al.* consider that a 'renewable revolution' can contribute to influence the energy markets in the following ten years, leading a deep change from oil and high costs of production to a renewable option (Loe *et al.*, 2014: 8–9).

Safety and environmental context

The new environmental international climate agreement at the United Nations Climate Change Conference, Paris 2015, along with diverse initiatives towards 2020 in relation to business activities in the Arctic, has been crucial to set the new framework for business. The environmental consequences of disasters in the Arctic are likely to be worse than in other regions while particular risk events – such as an oil-spill – are not necessarily more likely in the Arctic than in other extreme environments, and the potential environmental consequences, difficulty and cost of clean-up may be significantly greater, with implications for governments, businesses and the insurance industry. Trans-border risks, covering several national jurisdictions, add further complications (Emmerson and Lahn, 2012). Although the environment and people of the Arctic are transforming faster than anywhere else, this phenomenon is not experienced in the same way across the Arctic region (Arnason, 2007; Petrov, 2010 in Schmidt *et al.*, 2015).

An important aspect to highlight in the economic and environmental conditions in the Arctic is the considerable level of uncertainty. The main reason is that climate change affects all economic life activities.

There is a distinction between hazard and risk. On the one hand, hazard is anything with the potential to cause harm, such as substances, machines, work equipment, work activities, working environment and the effect of work organisation on human capital. On the other hand, 'risk considers both the probability of harm occurring, and the potential severity of that harm' (Renewable UK, 2014: 59).

Safety risk management may help to evaluate potential hazards before they occur and try to avoid negative consequences (Crain *et al.*, 2015: 5). The Arctic region has a broad variety of ecosystems that are very exposed to hazard and risks. Hence, the complex environment and the unprecedented disruptive and unpredictable transformations have challenged traditional local business. At the same time, this new scenario represents challenges and opportunities that will transform business in the Arctic (Loe *et al.*, 2014). Climate change associated with the warming of the sea and the melting of ice has impacted the industry, forcing it to adapt to the changeable conditions. The fast environmental changes experienced have favoured diverse analysis to study the future challenges for the industry (marine business, energy developments, communication, technology, economic growth, etc.) in the Arctic (Arctic Council Norwegian Chairmanship, 2009; Loe *et al.*, 2014; Perrals *et al.*, 2014, in Haavisto *et al.*, 2016).

Technology and innovation

The investment in reliable renewable energy technology based on safety systems and international standards is fundamental to meet the Paris Agreement goals. The business development in the Arctic confronts important technological challenges and infrastructure limitations. Large extensions of the Arctic are unavailable to develop commercial activities due to the lack of appropriate technology (Loe *et al.*, 2014: 9). At the national level, government incentives and support of creation of new policies in the development of green activities may promote investment in research into the development of new technologies to sustain environmentally friendly activities. Developing energy research capabilities and financing private investment may promote the advance of technology clusters (Loe *et al.*, 2014: 20). Also, public funds may encourage companies that green technologies may provide considerable business opportunities (Brun *et al.*, 2011).

According to the International Renewable Energy Agency (IRENA), in the last decade technological innovation contributed to reduce the cost of renewable power generation, and hydropower, biomass for power, onshore wind and geothermal may offer competitively priced electricity compared to fossil fuel-fired power generation (IRENA, 2015). The International Electro technical Commission (IEC), through technical committees (TCs), is in charge of producing International Standards for renewable energy systems to generate electricity from hydro, solar, marine, biofuels geothermal and wind energy. Different technologies have been developed to obtain energy from the sea, rivers and wave power. Marine energy technology production has interfered with traditional marine activities such as the fishing industry.

Regulations in the Arctic

The politics of Arctic economic development are contentious in the region with diverse perspectives among international, national and local levels, given the Arctic's iconic status and sensitive environment and strict global emission pathway (Haavisto *et al.*, 2016: 13). International cooperation supported by international treaties ensures that rules are set for the development, extraction and exploitation of natural sources, and establishing marine boundaries and shipping activities. However, energy resource extraction follows the national interests based on sovereignty principles and territorial boundaries (PAME, 2013; Andrew, 2014 in Haavisto *et al.*, 2016).

Governments play a central role in building policies to support the development of renewable generation plants (Brun *et al.*, 2011). Arctic governments have been developing an active role in order to 'set targets for increased energy efficiency, energy saving and the transition from fossil fuels to fuels with a low carbon footprint' (Jóhannesson, 2015: 17). For instance, the government in Iceland has played an active role in developing renewable resources in hydropower and geothermal energy.

Emmerson and Lahn (2012) have argued that there are major differences between regulatory regimes, standards and governance capacity across the Arctic States. The challenges of Arctic development demand coordinated responses where viable, common standards where possible, as well as transparency and best practice across the North. Governance frameworks need to be in place to allow sustainable development and support the public interest.

Large scale renewable energy generation

The Arctic represents a very complex challenge for companies to develop their activities. The environmental uncertainties motivate them to develop risk management systems to confront the challenges and manage their carbon and environmental footprint in the Arctic settings. The extraction and exploration in deep water may produce negative implications in the environment. The priority to reduce global greenhouse gas emissions is increasing the use of renewable energy. Emmerson and Lahn (2012: 5) have highlighted in the so-called Lloyds report that 'the businesses which will succeed will be those which take their responsibilities to the region's communities and environment seriously, working with other stakeholders to manage the wide range of Arctic risks and ensuring that future development is sustainable'.

The growing importance of global scale renewable energy

Diesel fuel has been a reliable source of energy for the communities in the Far North, however the risk associated with pollution, high cost and oil spills have motivated them to find a new source of energy. In this way, renewable energy companies have been growing in the North. The renewable energy companies that have invested successfully in the Arctic have identified their specific risks to develop the most adapted technology and services to the local conditions in order to thrive in the energy generation process. Also, Emmerson and Lahn (2012) have argued that those companies have to be aware that 'a long-term and comprehensive regulatory approach – incorporating national governments, bodies such as the Arctic Council, and industry bodies – is necessary for effective risk management, mandating cross-Arctic best practices and defining public policy priorities on what constitutes appropriate development' (Emmerson and Lahn, 2012: 18). Nevertheless, the prospects are substantial, the trajectory and speed of Arctic economic development are uncertain due to the unpredictable environmental conditions.

The main sources of renewable energy generation in the Arctic are hydroelectric and wind power, geothermal energy, solar photovoltaic and ocean energy. Hydroelectric power is the largest of the renewable energy sources widely used in the Arctic. Five of the Arctic States are in the top ten list of the world's largest hydropower generators: Canada (9.6 per cent), the United States (6.8 per cent), Russia (4.3 per cent), Norway (3.5 per cent) and Sweden (1.9 per cent) (IEA, 2017: 21). With regards to the percentage of total domestic electricity generation, Norway leads with 95.9 per cent, followed by Canada (56.8 per cent), Sweden (46.6 per cent), Russia (15.9 per cent) and the US (6.3 per cent). In Iceland, Finland and in Greenland hydropower also represents a significant share of the country's total electricity supply (IEA, 2017: 21–22). Hydropower and geothermal energy sources in Iceland have contributed to the country becoming the world's top aluminium producer (Loe *et al.*, 2014: 35).

In terms of global installed renewable energy capacity, wind power generation in the Arctic ranks second to hydropower and is expected to play a key role in the future of renewable energy development and climate change mitigation. In 2015, the net installed capacity of wind power in the Arctic region was weak. Except for Canada (11.2 per cent) and in the United States (4.5 per cent) (excluding Alaska), the installed capacity in most Arctic countries was insignificant (IEA, 2017).

Geothermal is a common energy source in the Arctic. Despite the fact that the United States leads in geothermal electricity generation, Iceland, Sweden and Norway are in front in terms of average annual use per person. Iceland's total electricity generation comes from geothermal resources. Thus, Iceland's hydroelectric and geothermal power allows it to produce minerals such as aluminium (Loe *et al.*, 2014).

Solar photovoltaic is one of today's fastest growing power generation technologies and is in use in all Arctic States except Iceland, where the preconditions for the extensive use of solar power are modest due to relatively low insolation. The potential for solar power to be used in Russia is great but progress has been slow. The largest installed capacity in the Arctic States is found in the United States (21.7 per cent) (IEA, 2017) and Canada. The contribution made by solar photovoltaic systems in terms of electricity supply in the other Nordic countries, including Greenland, remains minimal.

Renewable energy companies are increasingly looking to expand their investment opportunity in solar power installations in the Arctic. This is the case for private companies that have been increasing their operations in the Arctic. In Alaska, solar resource potential is similar to Germany, which has been the most important producer in photovoltaic capacity. While the Alaskan government is traditionally orientated towards fossil fuels, small private companies are taking the lead in promoting solar energy. But the state still receives scant attention from the large installation companies, most of whom have little or no presence in Alaska.

In the Arctic areas, Russia and Norway[2] have greater potential to generate ocean energy using tidal power small scale plants. However, only the coast of Alaska (Yukon, Eagle and Nenana rivers), Canada and Norway (Hammerfest) have the capacity to generate wave power among the Arctic States.

Principles and practice

Interpretations and attitudes towards risk are constructed based on diverse theoretical perspectives such as economics, social-psychology, culture and environmental sociology among others. From the sociological theory on risk and uncertainty, Ulrich Beck (1992) has argued that risk is inherent to modern societies and risk is not limited to one aspect such as environment or social patterns. The risk society is conceived by Beck as the 'world out of control' that creates manufactured uncertainties reinforced by accelerated technological innovation that promotes new global risks. The rapid technological change in areas such as atomic energy has shown uncontrollable consequences (Beck, 1992; Yates, 2001). Even though Becks' approach has a very relevant impact on risk research, the criticism is that his approach has not recognised the complexity of different aspects such as risk on governmental strategies (Dean, 1999) or different risk perceptions and responses (Tulloch and Lupton, 2003).

Understanding of the complex nature of risk is a relevant part to be considered but the human capability to control risk and change is a central aspect of risk and uncertainty studies. Along these lines, from Giddens' (1999) perspective, the emphasis on risk in modern societies is based on the belief that human beings can change societies for better. Thus, the future and security requires recognition of the need for risk management based on the idea that the future can be controlled or intervened in.

The concept of risk in the business environment has two contrasting meanings. On the one hand, risk is related to the assumption that everything is ruled by laws and can be known and understood as a result of the idea that everything can be measured and probabilities can be calculated. In this approach, probabilities are used to estimate risk. Even though events that occur regularly and probabilities offer a reasonable prospect of accuracy, there are diverse problems with the calculation that need to be taken into account (Harrison, 2014). Different variables such as technological change or misinterpretation about the probabilities can create problems in calculating of risk. Managerial approaches are, in general, based on an objective statistical and mathematical perception and a reductionist linear observation about risk can sometimes represent a limitation to organisations.

On the other hand, there is a contrasting perception of risk that recognises that there is not a single perception about risk and it cannot be known and controlled. In this way, Zhang (2011) proposed a framework against reductionism that integrates both subjective and objective approaches to risk management.

The sociological, environmental and integrative approach that incorporates elements from different theoretical perspectives on risk management may provide analytical elements to understand the complex uncertainties, rapid change and risk in the Arctic.

Risk in the Arctic

The Arctic Council declared in 1996 that oil and gas operations may have a potentially damaging environmental outcome. The Arctic Council drew clear parameters for suitable and successful processes for the protection of the Arctic environment through the Protection of the Arctic Marine Environment working group (PAME) (Arctic Council Norwegian Chairmanship, 2009). The Arctic Offshore Oil and Gas Guidelines represent a very important legal framework to establish the systems' safety management in order to avoid major disasters in Arctic offshore oil and gas production. Also, these guidelines set the rules for a common understanding of the process of managing risk.

The role of Arctic governments is to promote a safety culture in order to improve performance and protect the Arctic environment and natural resources. PAME developed a comparative analysis among the different country regulations and issued recommendations to improve safety in the offshore oil and gas industry and the marine environment (Arctic Council Nowergian Chairmanship, 2009). This analysis resulted in the identification of different types of risk in Arctic research: 'field safety risks for individuals (e.g. crevasse falls or exposure), risks for organisations (e.g. liability or personnel loss), risk to the environment (e.g. fuel spills or damage to tundra), and risks for science (e.g. equipment failure or data loss)' (Crain *et al.*, 2015: 5).

Uncertainty in the environment

Uncertainty is a central aspect of risk management: 'Managing risk requires thinking about risk, and thinking about risk requires thinking about and being comfortable with uncertainty and randomness' (Coleman, 2012: 2). The idea behind uncertainty is ambiguity, disorder and unknown environment. Risk management is not trying to change the situation but to transform or eliminate the uncertainties to different circumstances ruled by order and certainty. Risk management aims to control unavoidable uncertainties. In this way, Gigerenzer (2002) argues that sound statistical (and probabilistic) thinking can be enhanced through appropriate tools and techniques to approach uncertainty.

Uncertainty driven by politics is a central issue that may affect companies' operations in the Arctic countries. Macalister (2014) has pointed out that the complexity in developing large projects with different companies is that their activities can cross diverse jurisdictions areas. Additional constraints are that the eight Arctic countries do not have a unified legislation, and have different regulatory authorities.

With regards to risk management in energy projects, Irimia-Diéguez *et al.*, 2014 has identified in the literature of risk characterisation in megaprojects in the energy sector that the highest risk/uncertainty was mainly found in the construction risk, followed by customer/society risk and finally in the operation, design and financial risk. However, the main risk identified in the literature of energy in the Arctic has been focused on the effects of environmental uncertainties as the greatest risks to potential investors in Arctic economic development.

Sustainable and sufficient risk assessment in the Arctic

Assessing and managing Arctic risks

The Arctic has a complex risk environment. Many of the operational risks to Arctic economic development amplify one another: remoteness, cold and, in winter, darkness. The purpose of risk assessment is to identify the risks and guarantee that they are sufficiently assessed.

Jóhannsdóttir and Cook (2015: 19) have identified the main Arctic risks and they have classified in the different risks types. The main Arctic risks identified are: (a) operational factors: geographic isolation, lack of infrastructure and technology to confront disasters; (b) environment: oil spills, pollution, sea pollution during drilling, over fishing exploitation; (c) climate conditions: weather, storms, cold, darkness, icebergs; (d) regulations: litigation, national politics, different legal requirements, standards and governmental regulations, green and indigenous groups; (e) reputation: public opinion, contentious projects, companies' bad publicity.

The identification of the main risks shows a complicated environment for business to develop activities in the Arctic. Jóhannsdóttir and Cook (2015: 21) have argued that the Lloyd's report (Emmerson and Lahn, 2012) claims that the businesses operating successfully in the Arctic region will be those that take seriously their responsibility towards local communities and the environment, and which work with various stakeholders to manage their operations in a sustainable manner, aware of the risks and taking measures to mitigate them. These will be companies with comprehensive and rigorous risk management frameworks, enabling them to manage their own risk, use technologies and services most appropriate to Arctic conditions, and comply with all current and emerging legislative standards, such as the new Polar Shipping Code (Jóhannsdóttir and Cook, 2015: 21).

At the national level of governance, the need to coordinate relevant sectors and interests has often been managed through the creation of management plans, such as the Norwegian Management Plan for the Barents Sea, which integrates fisheries protection measures alongside the interests of oil and gas explorers (NME, 2011). When backed by international consultation and cross-cutting legislation, such plans could lead to the identification and protection of particularly sensitive replace regions across the Arctic region (Jóhannsdóttir and Cook, 2015: 21).

Risk management in the Arctic

Risk management includes a response and management of an identified risk. Risk assessment can be divided into a qualitative and a quantitative analysis. The methods of identifying risks include one-to-one meetings, brainstorming, nominal group techniques, the Delphi technique and qualitative evaluation methods (Jungeun *et al.*, 2016). Also, Dey has defined risk management as 'the systematic process of identifying, analysing and responding to project risk. It has six steps: planning, risk identification, qualitative risk analysis, quantitative risk analysis, risk response planning, risk monitoring and control' (Dey, 2012). Furthermore, 'it is an expanding field and literature has shown that it can be used not only for control against loss, but also as a way to attain greater rewards' (Dey, 2012; Wu and Olson, 2008 in Irimia-Diéguez *et al.*, 2014).

Hopkin (2017) has argued that risk management tools and techniques to manage hazard risk is the best and longest-established branch of risk management. Thus,

> hazard risks are associated with a source of latent harm or a situation with the potential to undermine objectives in a negative way and hazard risk management is concerned with mitigating the potential impact. Hazard risks are the most common risk associated with operational risk management, including occupational health and safety programmes.
>
> (Hopkin, 2017:18)

Risk management in megaprojects is one of the issues that has been less developed (Irimia-Diéguez *et al.*, 2014). Management of risk in complex projects is crucial due to the diverse parties that play an important role in the supply chain. Risk management is fundamental for companies to work safely, sustainably and successfully in the Arctic. Companies operating in the Arctic require robust risk management frameworks and processes that adopt best practice and take into account worst-case scenarios, crisis response plans and full scale exercises. There are many practical steps businesses can take to manage risks effectively, including investing in Arctic-specific technologies and implementing best-in-class operational and safety standards, as well as transferring some of the risks to specialist insurers (Emmerson and Lahn, 2012: 7). Risk management has a critical role to play in helping businesses, governments and communities to manage uncertainties and minimise risks. Also, it requires the latest data to analyse and control risks, especially those linked to environmental changes and global climate change, which in turn are giving rise to a broad set of economic and political developments.

Little (2011) has outlined that the aim of the risk management process (in every project size: small, medium or megaproject) is to identify and assess risks in order to enable the risks to be understood clearly and managed effectively. This process comprises six steps: planning, risk identification, qualitative risk analysis, quantitative risk analysis, risk response planning, and risk monitoring and control. Little has argued that after identifying the risk, a quantitative and qualitative risk analysis would be necessary. Finally, the risk needs to be planned, monitored and controlled. Risk management involves a variety of partners (public sector, construction companies, management companies, etc.). Hence, each risk needs to be analysed and then distributed, if possible, to various partnerships, which is the key to effective risk management. If risk can be transparently identified, equitably allocated and costed appropriately, successful projects will result (Little, 2011).

Models of risk assessment

The life-cycle model may aid understanding of complex uncertainties and in assessing risk in the Arctic. Following this approach, most of the risk cannot be addressed in isolation or understood as a lineal process. This model strategy is based on the assumption that the different elements play as entwined dimensions that interact with each other. This model strategy may be effective if diverse risks can be controlled (environmental, political, social, economic) simultaneously with enhancing risk management in the Arctic.

Life-cycle approach assessment

The main elements of the model are: identifying, assessing, reviewing, reporting and finally addressing risks. All these aspects of the model work together simultaneously to address risk. The elements of the life-cycle cannot be addressed in isolation. The life-cycle model has been presented as an alternative to be applied for the business to confront the uncertainties of the Arctic.

Life-cycle assessment (LCA) is increasingly used to assess production processes to support environmental decision-making and to give attention to the overall risk management situation both worldwide and in the Arctic. LCA is one of the most mature life-cycle-based methods for addressing problems related to environmental sustainability and it provides a framework to evaluate the environmental impacts of industrial activities in the Arctic region (Pettersen and Song, 2017). However, there are several weaknesses in the impact assessment methodology in LCA, e.g. related to uncertainties in the basic data used in LCA, such as Arctic seasonality and cold climate, as well as a more unpredictable environment, and this may potentially affect the impact assessment results. Other weaknesses include the absence of spatial differentiation and gaps in the coverage of impact pathways of different environments. Pettersen and Song (2017:14) have argued that Arctic-specific features and possible influences on LCA characterisation modelling need to be considered in the development of future Arctic life-cycle impact assessment. Pettersen and Song (2017) have also highlighted that life-cycle impact assessment for the Arctic is still very limited due to the unpredictable variability of the Arctic environment. However, LCA has been widely used for environmental analysis and as a decision tool for product development supporting government policy-making, industry and academia at continent, region and country levels. The risk assessment process can be used in offshore renewable energy installations (OREI) operations where the projects do not have historical data or the marine operations increase the risk such as the Arctic region.

Change in the energy industry and the implications for human capital

The change from fossil fuels to renewable energy has implications for human capital and how this is managed. The transformation of the industrial activity has repercussions for employment and how individuals are recruited to cover the business strategy.

Renewable energy and green jobs

The region has a wide variety of business such as mining, fishing, seafood, gas and oil, and these are the most relevant activities that have provided jobs to the local population. However, climate change has meant that these economic activities have had to be modified. Renewable energy production has created unprecedented employment conditions for the locals; new jobs, new technologies and people with different talents and expertise have arrived. The environment and the skilled manpower immigration have put an emphasis on the changing environment and the lack of knowledge of how to manage it has imposed many challenges. In this way most Arctic governments have provided incentives to create new green jobs. Government incentives include financial rewards or promote clear and stable energy policies in order to create a positive climate for investment in new technologies (Brun *et al.*, 2011).

Research institutes, non-governmental organisations and businesses can contribute to create green jobs, close the knowledge gap about environmental change and uncertainty, and to reduce risk, guaranteeing that development takes place within sensible ecological limits (Emmerson and Lahn, 2012).

Green jobs have been mostly defined as the kind of jobs that may contribute to look after the environment and which are frequently linked with environmental protection. The assumption that green jobs are safer because they help to improve biodiversity and reduce energy consumption it is not necessarily true. Workers in the green industry confront a series of hazards common and similar to any workplace, such as accidents or poor conditions. The United States Blue Green Alliance trade union refers to a green job as 'a blue-collar job with a green purpose' (Brun *et al.*, 2011).

Workers in green jobs have been exposed to new and different hazards that have not been broadly identified. For instance, the Occupational Safety and Health Administration (OSHA) in the United States has identified that workers in the solar energy industry confront a variety of health and safety hazards along the different stages of the production (OSHA, 2017a). Also, OSHA claims that workers may possibly be exposed to serious hazards, for instance carcinogenic substances such as Cadmium Telluride, if appropriate risk controls to eliminate hazards in the new industries are not in place. It is important that prevention of risks, e.g. through health and safety systems, are addressed as the solar energy industry grows.

The Occupational Safety and Health Administration's (OSHA) Electric Power Generation, Transmission and Distribution Standards require workers to implement health and safety workplace practices. Employers are required to give employees specific training and appropriate working conditions to prevent regular hazards and new risks inherent in green jobs. In the United States, the OSHA (2017a, 2017b, 2017c) standards cover most of the hazards in green industries and employers have to use essential controls to prevent possible risks in the workplace.

The change in the energy industry and in the renewable energy industry has impacted directly on the creation of new jobs. The International Labour Organization (ILO) has identified that the 'energy created through solar photovoltaic cells, landfill gas, or biomass plants have a higher number of jobs created per unit of energy produced than energy produced through conventional sources' (ILO, 2017). Job creation is a consequence of renewable energy diverse supply chains in production, manufacture and energy distribution.

The Renewable Energy and Jobs Annual Review 2017, (IRENA, 2017) has published:

> the renewable energy sector employed 9.8 million people, directly and indirectly, in 2016 – a 1.1 per cent increase over 2015. Jobs in renewable industry, excluding large hydropower, increased by 2.8 per cent to reach 8.3 million in 2016. Renewable energy employment worldwide has continued to grow since IRENA's first annual assessment in 2012, but the last two years have seen a more moderate rate of growth. The most consistent increase has come from jobs in the solar PV and wind categories, together more than doubling since 2012. In contrast, employment in solar heating and cooling and large hydropower has declined.
>
> (IRENA, 2017)

Conclusion

The future of large scale renewable energy generation in the Arctic region has led to the establishment of safety procedures through the risk management process. The renewable energy business that will succeed in the Arctic will be the one that takes a management risk strategy incorporating local and transborder Arctic legislation, environmental conditions, local communities' responsibilities, specific employment circumstances, and manages the changeable complex uncertainties in order to guarantee sustainable development in the Arctic region.

The assumption based on the idea that the future can be controlled or influenced through the traditional linear management paradigm would be too restrictive to approach uncertainty and complex risk problems. In this way, an integrative approach that constructs a framework integrating subject and object approaches to risk management (Zhang, 2011) may help understanding the complexity of risk and the rapidly changing conditions and specific uncertainties in the renewable energy and business environment in the Arctic.

Notes

1 Renewables includes 'hydro, geothermal, solar PV, solar thermal, tide/wave/ocean, wind, municipal waste (renewable), primary solid biofuels, biogases, biogasoline,

biodiesel, other liquid biofuels, non-specified primary biofuels and waste and charcoal' (IEA, 2017: 73).

2 Norway launched the world's first wave power station at Berger (World Energy Council (WEC), 2016: 29).

References

Andrew, R. (2014) *Socio-Economic Drivers of Change in the Arctic.* Oslo, Norway: Arctic Monitoring and Assessment Program (AMAP).

Arctic Council. (2017) *The Arctic Council: A Backgrounder*, www.arctic-council.org/ index.php/en/, accessed 11 November 2017.

Arctic Council Norwegian Chairmanship. (2009) *Arctic Marine Shipping, Assessment 2009 Report*, PAME, www.pmel.noaa.gov/arctic-zone/detect/documents/AMSA_2009_ Report_2nd_print.pdf, accessed 9 October 2017.

Arnason, R. (2007) 'Climate change and fisheries: assessing the economic impact in Iceland and Greenland', *Natural Resource Modelling*, vol. 20, no. 2, pp. 163–197.

Beck, U. (1992) *Risk Society: Towards a New Modernity*, London, Newbury Park, Sage.

Brun, E., Ellwood, P., Bradbrook. S. and Reynolds, J. (2011) *SAMI Consulting Foresight of New and Emerging Risks to Occupational Safety and Health Associated with New Technologies in Green Jobs by 2020, Occupational, Safety and Health Association*, https://osha.europa.eu/en/tools-and-publications/publications/reports/foresight-green-jobs-drivers-change-TERO11001ENN, accessed 11 November 2017.

Coleman, T. (2012) *Quantitative Risk Management: A Practical Guide to Financial Risk*, Hoboken, New Jersey, John Wiley & Sons.

Crain, R., Creek, K. and Wiggins, H. (eds) (2015) *Supporting a Culture of Safety in Arctic Science Workshop Report: Arctic Field Safety Risk Management*, https://commssite.files.wordpress.com/2013/09/nsf_srm_report_2015_reducedsize.pdf, accessed 14 November 2017.

Dean, M. (1999) 'Ordering risk', in D. Lupton (ed.) Risk and Sociocultural Theory: New Directions and Perspectives, pp. 168–185, Cambridge, Cambridge University Press.

Dey, P. K. (2012) 'Project risk management using multiple criteria decision-making technique and decision tree analysis: a case study of Indian oil refinery', *Production Planning & Control*, vol. 23, no. 12, pp. 903–921.

Emmerson, Ch. and Lahn, G. (2012) *Arctic Opening: Opportunity and Risk in the High North*, Chatman House, Lloyds, file:///C:/Users/User/Downloads/Arctic_Risk_Report_ webview%20(1).pdf, accessed 4 November 2017.

Giddens, A. (1999) 'Risk and responsibility', *Modern Law Review*, vol. 62, no. 1, pp. 1–10.

Gigerenzer, G. (2002) *Calculated Risks: How to Know When Numbers Deceive You*, New York, Simon and Schuster.

Glomsrød, S. and Aslaksen, I. (eds) (2009) 'The Economy of the North 2008', *Statistisk Sentralbyrå/Statistics Norway*, www.ssb.no/a/publikasjoner/pdf/sa112_en/sa112_ en.pdf, accessed 27 November 2017.

Haavisto, R., Pilli-Sihvola, K., Harjanne, A. and Perrels, A. (2016) 'Socio-economic scenarios for the Eurasian Arctic by 2040', *Finnish Meteorological Institute*, https:// helda.helsinki.fi/bitstream/handle/10138/160254/2016nro1.pdf, accessed 11 October 2017.

Harrison, A. (2014) *Business Environment in a Global Context*, Oxford, Oxford University Press.

Hopkin, P. (2017) *Fundamentals of Risk Management Understanding, Evaluating and Implementing Effective Risk Management*, 4th edition, London, Kogan Page.

International Energy Agency (IEA). (2017) *Key World Energy Statistics*, www.iea.org/publications/freepublications/publication/KeyWorld2017.pdf, accessed 29 October 2017.

International Labour Organisation (ILO). (2017) *Green Jobs and Renewable Energy: Low Carbon, High Employment*, www.ilo.org/wcmsp5/groups/public/-ed_emp/-emp_ent/documents/publication/wcms_250690.pdf, accessed 13 November 2017.

International Renewable Energy Agency (IRENA). (2015) *Renewable Generation Costs in 2014*, www.irena.org/documentdownloads/publications/irena_re_power_costs_2014_report.pdf, accessed 30 November 2017.

International Renewable Energy Agency (IRENA). (2017) *Renewable Energy and Jobs Annual Review 2017*, www.irena.org/DocumentDownloads/Publications/IRENA_RE_Jobs_Annual_Review_2017.pdf, accessed 20 December 2017.

Irimia-Diéguez, A., Sánchez-Cazorla, Á. and Alfalla-Luque, R. (2014) Risk management in megaprojects, *Social and Behavioral Sciences*, vol. 119, pp. 407–416.

Jóhannesson, G. (2015) 'Iceland – Renewable as a National Project in Renewable Energy in the Arctic', *The Circle*, WWF Global Programme, file:///C:/Users/User/Downloads/thecircle0315_web.pdf, accessed 11 October 2017.

Jóhannsdóttir, L. and Cook, D. (2015) *An Insurance Perspective on Arctic Opportunities and Risks: Hydrocarbon Exploration & Shipping, Iceland*, Centre for Arctic Policy Studies/Institute of International Affairs/University of Iceland, Working papers, http://ams.hi.is/wp-content/uploads/2015/04/An_Insurance_Perspective_PDF.pdf, accessed 21 October 2017.

Jungeun P., Park, B., Cha, Y. and Hyun, C. (2016) Risk factors assessment considering change degree for mega-projects, *Procedia – Social and Behavioral Sciences*, no. 218, pp. 50–55.

Little, R. G. (2011) The emerging role of public-private partnerships in megaproject delivery, *Public Works Management and Policy*, vol. 16, no. 3, pp. 240–249.

Loe, J., Fjærtoft, D. B., Swanson, P. and Jakobsen, E. W. (2014) 'Arctic Business Scenarios 2020. Oil in Demand, Green Transformation Refreeze', Norway, *Menon Business Economics-Arctic Business*, file:///C:/Users/User/Downloads/Arctic+Business+Scenarios+2020+pages.pdf, accessed 2 October 2017.

Macalister, T. (2014) 'Shell's Arctic drilling set back by US court ruling: appeal court rules environmental risks have not been properly assessed by government in victory for green groups', *Guardian*, 23 January 2014.

Norwegian Ministry of the Environment (NME) (2011) First Update of the Integrated Management Plan for the Marine Environment of the Barents Sea-Lofoten Area. Report to the Storting (White Paper). Recommendation of 11 March 2011 from the Ministry of the Environment, Approved in the Council of State the Same Day, Oslo, NME.

Occupational Safety and Health Administration (OSHA). (2017a) *Green Jobs Hazards: Solar Energy*, United States, Department of Labour, www.osha.gov/dep/greenjobs/solar.html, accessed 11 December 2017.

Occupational Safety and Health Administration (OSHA). (2017b) *Green Hazards Geothermal Energy*, United States, Department of Labour, www.osha.gov/dep/greenjobs/geothermal.html, accessed 11 December 2017.

Occupational Safety and Health Administration (OSHA). (2017c) *Green Hazards*, United States, Department of Labour, www.osha.gov/dep/greenjobs/, accessed 15 December 2017.

PAME (2013) *The Arctic Ocean Review Project, Final Report*. Kiruna: Protection of the Arctic Marine Environment (PAME) Secretariat.

Perrels, A., Prettenhaler, F., Kortschak, D., Heyndrickx, C., Ciari, F., Bösch, P. and Thompson, A. (2014) 'Sectoral and cross-cutting multi-sector adaptation strategies for energy, transport and tourism'. FP7 ToPDad Project.

Petrov, A. (2010) 'Post-sample bust modelling economic effects of mine closures and post-mine demographic shifts in an Arctic economy (Yukon)', *Polar Geography*, vol. 33, no. 1 and 2, pp. 39–61, https://doi.org/10.1080/1088937X.2010.494850, accessed 9 October 2017.

Pettersen, J. B. and Song, X. (2017) 'Life cycle impact assessment in the Arctic: challenges and research needs', *Sustainability Journal*, vol. 9, no. 9, file:///C:/Users/User/Downloads/sustainability-09–01605-v2%20(1).pdf, accessed 18 December 2017.

Renewable UK. (2014) *Offshore Wind and Marine Energy Health and Safety Guidelines*, issue 2, http://c.ymcdn.com/sites/www.renewableuk.com/resource/collection/AE19ECA8-5B2B-4AB5-96C7-ECF3F0462F75/Offshore_Marine_HealthSafety_Guidelines.pdf, accessed 18 December 2017.

Schmidt, J., Aanesen, M., Klokov, K. B., Khrutschev, S. and Hausner, V. H. (2015) 'Demographic and economic disparities among Arctic regions', *Polar Geography*, vol. 38, no. 4, https://doi.org/10.1080/1088937X.2015.1065926, accessed 20 December 2017.

Tulloch, J. and Lupton, D. (2003) *Risk and Everyday Life*, London, Sage.

World Energy Council (WEC). (2016) 'Marine Energy 2016', *World Energy Sources*, www.worldenergy.org/wp-content/uploads/2017/03/WEResources_Marine_2016.pdf, accessed 22 December 2017.

Wu, D. and Olson, D. L. (2008) Supply chain risk, simulation, and vendor selection, *International Journal of Production Economics*, vol. 114, no. 2, pp. 646–655

Yates, J. (2001) 'An Interview with Ulrich Beck on Fear and Risk Society', *The Hedgehog Review*, University of Virginia, Institute for Advanced Studies in Culture, www.iasc-culture.org/THR/archives/Fear/5.3HBeck.pdf, accessed 23 December 2017.

Zhang, H. (2011) 'Two schools of risk analysis. A review of past research on project risk', *Project Management Journal*, vol. 42, no. 4, pp. 5–15.

Part VI

Arctic energy and non-Arctic world

11 The case for increased UK–Nordic electricity interconnection and the implications of Brexit

Nicholas Craig

Introduction and background

As the UK steps away politically from the European Union (EU) with one foot, it steps towards it with the other, through the ambition of greater electricity interconnection. Increasing cross-border interconnection is a key pillar of both the UK and the EU's energy and climate policy, and offers significant socioeconomic benefits (BEIS, 2017a). By increasing the security of supply, reducing consumer costs and allowing greater exploitation of renewable resources, interconnectors have been earmarked as a crucial technology in the energy transition towards a low-carbon, flexible electricity system (Moore, 2014).

Energy resources in the Nordic and Arctic region have historically been of economic and strategic importance to the UK. As well as the oil and gas industry in the UK's North Sea, British interests also extend further towards the Arctic Circle. Following a spark of interest in the past decade in the potential for untapped hydrocarbon resources in the Arctic, British firms have been behind much of the exploratory work in the region, with involvement in seven out of the ten drilling licences issued by Norway in 2016 (Craig *et al.*, 2017).

However, the development of hydrocarbons in the Arctic is not only costly, but also has a number of risks attached. Given the significant challenges and cost of hydrocarbon extraction in the Arctic's harsh conditions and an increasing number of legal battles, energy companies are seeking to limit the risk of stranded assets in the region. Moreover, the volatility of oil prices and the potential economic, environmental and reputational risk of an accident in the pristine Arctic environment has effectively put an end to exploration in high-risk areas such as off the coast of Greenland and North America (Craig *et al.*, 2017). There is also a growing body of research suggesting that development of Arctic hydrocarbons is incommensurate with the goal of the Paris Agreement, to keep global warming well below 2°C (McGlade and Ekins, 2015).

Given these problems with Arctic hydrocarbon extraction, a new narrative is needed on which to base future energy relations between the UK and its Nordic neighbours.

As the UK transitions to a low-carbon energy system, where intermittent renewable resources will play an increasing role, there is a growing need for interconnection to offer flexibility. This chapter will examine the role of inter-connectors in Britain's low-carbon transition, with a particular focus on greater interconnection to Nordic nations. Benefits and risks of increased interconnection will be put forward, followed by a case-by-case analysis of interconnection between the UK, Denmark, Norway and Iceland. Finally, the implications of Britain's impending exit from the EU for cross-border interconnection will be discussed.

The UK's energy transition

In 1882, the world's first coal-fired power station opened in London. The UK's rapid economic growth through the industrial revolution and path to global power in the twentieth century was fuelled by coal. Fast forward to today, and the UK government is now committed to phasing out all unabated coal use by 2025, and leading the international 'Powering Past Coal' coalition (BEIS, 2018). While the UK is the birthplace of coal power, it is now at the forefront of making it obsolete.

This move signals the UK's shift to a low-carbon energy system, a pathway set out in law through the UK's 2008 Climate Change Act. The ambition laid out in the Clean Growth Strategy is of an 80 per cent reduction in power emissions by 2032 from 1990 levels, driven primarily by an increase in renewable generation (BEIS, 2017b). By mid-2017, the UK had 38 GW of installed renewable capacity, of which 79 per cent is solar and wind (BEIS, 2017a). These two technologies are inherently intermittent, producing electricity only when the sun shines or the wind blows. As such, a future energy system that is increasingly reliant on wind and solar needs growing flexibility to balance this intermittency.

The capacity market is one of the UK's primary mechanisms to ensure ample supply at times of system stress. Currently however, fossil fuel generation accounts for much of that, with over 50 per cent of 2017's capacity auctioned to fossil fuel power plants (Ward, 2017). This means that the UK government is in effect subsidising fossil fuels through this mechanism, slowing the low-carbon transition. Increasing interconnection provides an effective way to change this outcome.

UK interconnection

Interconnection is a proven and cost-efficient way to increase flexibility within the power system. It is identified by the National Infrastructure Commission as being a leading innovation to 'fire a smart power revolution' alongside storage and demand response (National Infrastructure Commission, 2016, p. 4). Interconnectors allow electricity to be moved from markets with low cost or high availability to those where costs or demand is higher, meaning that electricity can be imported during peak demand, and exported when generation exceeds demand. The UK currently imports a great deal more power than it exports: 100 TWh imported, for every 16 TWh exported in the year to March 2017 (Lodge and Mahoney, 2017).

The UK currently has 4 GW of interconnection, to France, Ireland and the Netherlands: 2 GW short of the 10 per cent EU goal by 2020. However, another 4.4 GW is under construction and a further 9.5 GW is expected to come online by the mid-2020s, connecting the UK to Belgium, Norway, Denmark, and pushing the UK over the 2030 EU goal of 15 per cent interconnection (BEIS, 2017b). Some analysis has suggested that up to 35 GW of interconnection would be beneficial for the UK and Europe as a whole (European Climate Foundation, 2011). Figure 11.1 shows the UK's existing and proposed interconnectors.

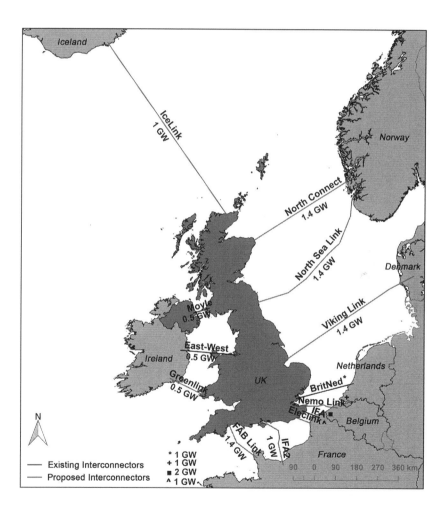

Figure 11.1 Map of existing and planned UK interconnectors.

Source: Kirby, 2018 – adapted from Vaughan, 2017 for this book chapter; capacity data from Ofgem, 2017c.

Interconnectors can cut costs and carbon

Increased interconnection has the potential to reduce the cost and carbon intensity of the UK's electricity supply. As carbon is increasingly attributed an economic cost through mechanisms such as the UK's carbon floor price, it is possible to calculate the potential value of carbon savings through increased interconnection. Based on a £30/tCO$_2$ price, each additional GW of interconnection can generate carbon savings worth £1.5–£115 million annually (Moore, 2014).

Interconnectors improve market efficiency, meaning there is a strong economic case for increased interconnection for both consumers and generators. Electricity delivered through interconnection, even at long range, is cost-competitive with other forms of power generation in the UK, and negates the need to build expensive new generation capacity (Moore, 2014). On this basis, a doubling of interconnection capacity has the potential to reduce consumer costs by £1 billion annually by 2020, with each additional GW of interconnection reducing wholesale energy costs by 1–2 per cent, (National Grid, 2014; Mount *et al.*, 2016).

As the proportion of intermittent generation increases, price volatility is predicted to increase, as demonstrated by negative electricity prices recently reached in Germany during periods of high wind (Reed, 2017). Interconnection can stabilise prices for consumers, as well as offer greater opportunities for renewable power generators to continue to generate and export power when domestic demand is low.

One of the problems associated with increased wind generation is the inability to match production to demand. This results in new system balancing costs for National Grid in the form of constraint payments to large wind farms to stop generating in times where supply outstrips demand, or bottlenecks occur in the transmission system. The cost of constraint payments for wind, which is passed onto consumers, exceeded £108 million in 2017 (Renewable Energy Foundation, 2018). Increasing interconnection, however, would prove beneficial for the system here, as planned interconnectors could halve the amount of curtailed renewable electricity generation in the UK to 15.1 TWh annually, saving the consumer significant sums (Strbac *et al.*, 2012).

The UK's current policy approach

Given the associated positive benefits, the UK is pursuing greater levels of interconnection, with the Secretary of State, Greg Clark, stating in April of 2017 that 'the ambition is to go higher' (Vaughan, 2017). Several policy interventions have introduced mechanisms – such as the 'cap and floor' regime – that enable private actors to initiate economically viable interconnector projects, leaving market forces to set the future direction.

In 2014, the UK introduced the 'cap and floor' regime for interconnector projects that are assessed to be of potential benefit to the consumer. The scheme reduces financial risks for developers with a guaranteed minimum 'floor' price for electricity, while ensuring value for the consumer by setting a price 'cap'. While interconnectors are now included in the UK's capacity market, longer contract lengths will be needed to allow these projects to submit more competitive bids (Craig *et al.*, 2017).

Risks of greater interconnection

While interconnection brings many benefits, there are also associated risks. Interconnectors provide competition to domestic suppliers, and the UK's current policy framework creates an uneven playing field. Electricity from interconnectors is currently exempt from certain levies and charges, including transmission fees and the carbon floor price, giving foreign generators an estimated £10/MWh advantage (Aurora Energy, 2016). While it remains agreed that consumer prices are lowered through interconnection, some argue that these policy flaws are enough to introduce wider net negative economic impacts and revenue reductions of up to 10 per cent for domestic generators (Aurora Energy, 2016).

Exemption from the carbon floor price is argued to create inadvertent incentives for foreign generators to sell electricity of higher carbon intensity to the UK. This has been witnessed in the case of the BritNed interconnector, where the import of Dutch electricity, generated mostly with fossil fuels, resulted in a 30 per cent higher carbon intensity of imported electricity than domestic generation in the year preceding October 2017 (Staffell *et al.*, 2017).

Critics contend that increasing interconnection is negative for the UK's energy security, suggesting that investment would be better placed in storage capacity (Lodge and Mahoney, 2017). However, they do concede that the benefits differ from interconnector to interconnector (Aurora Energy, 2016).

Why UK–Nordic interconnectors?

While the case for greater interconnection is strong, it is also clear that not all interconnectors offer the same benefits, and many have greater risks attached, as demonstrated with high-carbon imports through the Netherlands link. The UK's National Infrastructure Commission signals that focus should be given to expanding interconnection with markets that are currently poorly connected, where prices, demand patterns, energy mixes and carbon intensities differ significantly to the UK (National Infrastructure Commission, 2016).

Examining the carbon intensity and electricity price data from the UK and potential interconnected markets in Figure 11.2, it is clear that the UK's Nordic neighbours Denmark, Norway and Iceland, offer several advantages in terms of potential economic and environmental benefit compared to continental Europe. Combined with the lack of existing interconnection to these systems, it is no surprise that the UK has begun to look to connect to its Nordic neighbours. With complementary and often more dispatchable energy mixes and low-carbon intensities in the Nordics, these cables have particularly high structural values (Craig *et al.*, 2017).

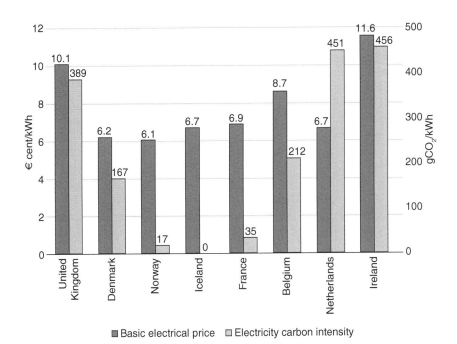

Figure 11.2 Non-domestic electricity price, without taxes and levies (2016, € cent/kWh) and carbon intensity of electricity generation (2014, g CO^2/kWh) in potential interconnected markets.

Sources: European Environment Agency, 2016; Norwegian Water Resources and Energy Directorate, 2016; Eurostat, 2017.

There are currently four interconnectors at various stages of construction or planning between the UK and the Nordics: VikingLink (Denmark), North Sea Link and NorthConnect (Norway) and IceLink (Iceland). As will be explored further, each of these Nordic energy systems presents a unique opportunity for symbiosis with the UK's electricity network.

Examining Figure 11.3, it is clear that the capital cost involved in connecting UK and Nordic markets is far higher than closer markets. The cost of Viking Link for example, is almost twice that of Nemo, and IceLink would require almost five times the capital as ElecLink. Connecting these networks also involves pushing the boundaries of interconnector lengths to new records. As with all transmission of electricity, there are thermal losses of electricity both in conversion and transmission, and based on research by Pöyry (2016), UK–Nordic interconnection would introduce loss factors of 7.8 to 11.3 per cent compared to ~3 per cent for French and Belgium links. With such long cables, there is also a greater scope for technical difficulties to arise, although as they occur randomly, the impact of such failures on the socio-economic value of the projects are difficult to assess (National Infrastructure Commission, 2016).

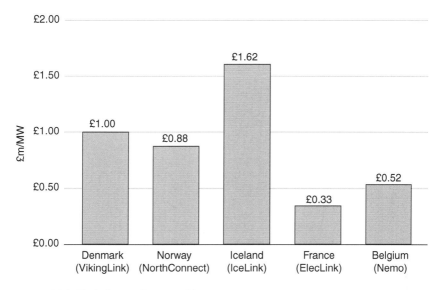

Figure 11.3 Capital cost of proposed interconnector per MW.

Source: adapted from Moore, 2014.

The following sections will explore in detail the particular benefits and risks of increasing interconnection between the UK and Norway, Denmark and Iceland. For VikingLink, the synchronicity of two increasingly wind-heavy energy systems and Danish criticism of the project will be examined. For Norway, the low-carbon potential, and the ability of pumped storage in Norway to play an energy storage role in the UK system shall be considered, and looking further North to Iceland, the non-standard elements of this interconnector which make its future more uncertain shall be highlighted.

The UK and Denmark

The 770 km long, 1.4 GW VikingLink interconnector is currently in the planning process and is expected to come online in 2022 to connect the Danish and UK electricity markets. In October 2017, the Danish Energy Minister confirmed an 11 billion Danish Kroner (DKK) investment as part of a wider push to increase Danish interconnection with the UK, Germany and the Netherlands. This has been seen as a positive signal for the pending final investment decision to be made in March 2018.

Linking the UK and Danish electricity networks offers socio-economic benefits on both sides, and as such has been listed on the EU's Projects of Common Interest (PCIs) as particularly important for developing the European Internal Energy Market (IEM). The capital cost of the project is estimated to be £1.4 billion, similar to the cost of connecting to Norway (see Figure 11.3). Cost-benefit analyses of

VikingLink estimates a net economic gain of 4.7 billion DKK (~£561 million) for the Danish economy by 2040 and a benefit of £0.9 to £1.8 billion for UK consumers (Pöyry, 2014; Energinet, 2017).

Both countries have significant wind resources, with each looking to further exploit offshore potential in the future. In some recent cases, Denmark has produced 140 per cent of its demand from wind power (Nelsen, 2015). By 2020, an estimated 50 per cent of Danish electricity will be produced using wind power, with a government target of total reliance on renewables by 2050 (Danish Energy Agency, 2017).

With increasing reliance on wind power, the synchronicity of wind patterns between the two countries is important to understand when assessing the potential benefits of interconnection.

Wind power production between the UK and Denmark has a synchronicity of 0.37 (where 0 indicates two systems that never simultaneously experience the same wind production patterns) (Monforti *et al.*, 2016). Such poor correlation between wind production in the two countries indicates the possible benefits from connecting the systems, as they rarely experience the same patterns of wind production. When compared to the synchronicity between the UK and other potentially connected markets, such as the Netherlands (0.63 correlation) and Belgium (0.60 correlation), it is clear that from the perspective of increased intermittent wind production, a UK–Denmark connection would be more beneficial.

The weather patterns that bring wind to both the UK and Denmark offer an explanation for this. As both countries lie between 50 and 60 degrees latitude, they are both affected by prevailing westerly and south-westerly winds, meaning that winds typically arrive in the UK approximately 24–48 hours before Denmark (Met Office, 2017; Miljøstyrelsen, 2017). This indicates a strong potential for balancing between the two systems over such timescales, as was demonstrated during the October 2017 Storm Ophelia, where peaks in wind production in the UK were approximately one day ahead of Denmark.

While there are clear advantages to VikingLink, particularly with the synchronicity of the wind-powered systems, there has also been criticism over the economics from a Danish perspective. The 11 billion DKK investment is set to be paid from Danish consumers' energy bills, with prices estimated to increase by 8–40 DKK annually for households, and up to 100,000 DKK for medium-sized firms (Bjerregaard, 2017). Analysis has suggested that while the Danish government's estimated 4.7 billion DKK gain by 2040 may be robust, greater returns could be achieved by instead investing in heat pumps to store excess energy domestically (Ea Energianalyse, 2015).

A further point worth exploring is that of the carbon intensity of marginal electricity production, which could theoretically be the electricity exported if VikingLink created higher demand on the Danish system. Denmark's marginal capacity today consists of imported power, comprising primarily of coal and gas generation, with a carbon intensity of $875\,g\,CO_2/kWh$, five times higher than the figure shown in Figure 11.2 (Ea Energianalyse, 2016). This has fuelled the argument that the UK could, through VikingLink, end up importing high-carbon German power through the cable.

However, with the long life-time of an interconnector and the rapid low-carbon transition taking place across Europe, it is more important to consider long-term marginal capacity (Lund *et al.*, 2010). Some forecasts have suggested that by 2020, marginal capacity in Denmark will be entirely renewable, with 80 per cent coming from wind power (Muñoz, 2015). If true, that would negate the argument of carbon imports through this interconnector, but the debate demonstrates the complexity of predicting the dynamic energy markets of the future.

The UK and Norway

Two projects, North Sea Link and NorthConnect, are currently planned to link the British and Norwegian electricity markets. The 1.4 GW, 720 km long North Sea Link, is already under construction and is expected online by 2022 (Ofgem, 2017b). The second is the NorthConnect project which has recently been announced as in the interest of British consumers, and likely to be awarded regulatory approval imminently (Ofgem, 2017a).

As shown in Figure 11.2, the Norwegian electricity network has an extremely low-carbon intensity of just 17 g CO_2/kWh, compared to 389 g CO_2/kWh in the UK. Connecting the two markets through the two proposed interconnectors would require a capital cost of £0.88–£1.18 million per MW, less than a third the cost of the UK government's central estimates for new nuclear generation (Moore, 2014).

The Norwegian government's policy approach to further develop renewable resources are particularly focused around wind and hydropower, and specify that interconnection is a key element in the future electricity system. New legislation in Norway has been aimed at liberalising the interconnection market, allowing private actors, not just state-owned entities, to build new links (Norwegian Ministry of Petroleum and Energy, 2016).

Ninety-seven per cent of Norway's electricity production is generated from hydropower. This provides particularly noteworthy elements to consider when discussing interconnection with the UK. First, hydropower is a highly dispatchable resource that can start producing electricity at very short notice. This is attractive in terms of adding flexibility, allowing the UK to buy electricity ahead of time, when low winds or solar radiation are forecast. Contrasted to a UK–Denmark connection, a Norwegian link would provide more stable balancing potential. While hydropower can be vulnerable to periods of low rainfall over seasonal timescales, it generally has an availability of 78–90 per cent (Ofgem, 2013).

Norway's access to abundant hydropower resources also means a great potential for pumped storage. The country currently has 1.4 GW of pumped storage capacity, and studies have suggested that up to 20 GW additional capacity could be added (Grøv *et al.*, 2011; IHA, 2017). Utilising this, the UK could export renewable electricity to Norway at times of high supply and store it, before importing it back when demand is higher. This would provide a use for excess

UK wind energy and reduce the need for curtailing and constraint payments. With 93–99 per cent availability, pumped storage has higher availability than any other generation source including coal, gas or nuclear (Ofgem, 2013).

Using the cable as a connection to a giant Norwegian 'battery' must factor in the high cost of the interconnector and the transmission losses of ~8 per cent each way. However, such losses compare favourably to alternatives, for example in the charge and discharge of grid-scale battery storage. There has also been research that draws into question the ability of Norway to scale up its pumped storage, the environmental impact of doing so and its own domestic need for a more flexible electric grid (Deign, 2017). Economic analysis by Henden *et al.* (2016) suggests an optimal of just 0.95 GW extra pumped storage capacity, and with other nearby countries also looking to connect to the same storage, it is not a simple picture. In any case, it is still touted by proponents of a UK–Norway interconnector as a potentially interesting benefit (Grøv *et al.* 2011).

Particularly important when considering the ability of Norwegian interconnectors to lower carbon emissions is Norway's carbon intensity of marginal electricity: $0.00\,g\,CO_2/kWh$. Compared to other interconnected markets such as France and Belgium which have similar carbon intensities to the UK at the margin, this indicates that importing electricity from Norway into the UK can deliver significantly reduced carbon emissions in a cost-effective manner (Moore, 2014). At a carbon price of £30/tCO$_2$, this would represent an estimated saving to the UK of £115 million per year, at a cost of £17 per tonne of carbon saved, which is less than a third of the cost that offshore wind or nuclear can deliver (Moore, 2014).

The UK and Iceland

Of all the interconnectors in Figure 11.1, IceLink, the proposed interconnector between the UK and Iceland is by far the most ambitious and complex, but offers a great deal of potential benefits. Such an idea is hardly new, as in 1988, *The Economist* laid out the potential gains for the UK and Iceland from interconnection (*The Economist*, 1988).

That same article cited power in Iceland as 'the cheapest in Europe', stating that 'Icelandic watts would be cheaper than those from the still unbuilt nuclear flagship' (*The Economist*, 1988). Thirty years later, Iceland's electricity prices remain low, and given the UK government's pursuit of Hinkley Point, branded the world's most expensive nuclear power plant, the comparisons are easy to draw (Watt, 2017). In fact, the £1.6 million per MW capital cost of IceLink (Figure 11.3), pales in comparison to estimates for Hinkley Point of £6.3 million per MW (Watt, 2017). This was a point raised by Angus MacNeil, former Chair of the Committee on Energy and Climate Change, at the 2016 Arctic Circle Assembly, where he outlined the strategic, security and economic benefits of IceLink.

Iceland's electricity system is entirely renewable, being delivered by hydro and geothermal resources. This allows it to be the only potential interconnected market to the UK where carbon intensity of electricity, as shown in Figure 11.2, is zero. This mix is vastly different to the UK's, and represents a good balance of baseload and dispatchable power generation at a low price.

However, more than any other interconnector, this would be an export cable, with flow from Iceland to the UK 85 per cent of time (Pöyry, 2016). Given a current lack of excess generating capacity in Iceland, this means that the link would be built on the principle of increased generation to meet the extra 1 GW demand. Concern has been expressed in Iceland that exporting power would mean importing prices, and assessments suggest that upgrades to the countries' transmission system would increase costs up to £2.7 million per MW capital cost (Pöyry, 2016; Neslen, 2017). Furthermore, electricity pricing mechanisms in Iceland are not well aligned to the UK, presenting additional challenges.

While the concept of such a cable goes back almost 60 years, it is only within the last decade that it has been considered economically feasible (Landsvirkjun, 2017). In 2015, bilateral talks ended with the establishment of a task-force, which found that despite unaddressed issues, the project would deliver significant socio-economic benefits for both sides (Neslen, 2017). In 2017, this was bolstered by new research in Iceland that has the potential to increase geothermal output from wells by a factor of 10, increasing the profitability of power production in the country (Karagiannopoulos, 2017). Both countries are exploring the next steps of feasibility studies and preparatory work, and the project has been placed on the EU's list of PCIs (Landsvirkjun, 2017).

Brexit

Since voting to leave the EU in June 2016, the UK has begun the process of withdrawal and is expected to have left the union by April 2019. At the time of writing, there remains great ambiguity on the future relationship between the UK and the EU.

The EU's energy policy strategy, to ensure 'secure, affordable and climate-friendly energy' through an increasingly interlinked and liberalised energy market has historically been supported and often spearheaded by the UK (Froggatt *et al.*, 2017). Post-Brexit, such goals are likely to remain well aligned, with the UK's domestic legislation underpinning the government's commitment to delivering clean growth. In order to rapidly decarbonise at the lowest possible cost to taxpayers, businesses and consumers, high levels of UK–EU cooperation on energy will be needed post-Brexit (Kumar, 2017).

Speaking in November 2016, the UK's Secretary of State for Energy and Climate Change said:

> Brexit does not change the advantages of linking our electricity market with those of Europe.
>
> (Clark, 2016)

While he is correct in that many of the fundamental benefits of increased interconnection laid out in this chapter still exist post-Brexit, the UK's departure from the EU also introduces a great deal of uncertainty, as well as several risks in both the short and long term. Of the four UK–Nordic interconnectors examined in this chapter, only IceLink and NorthConnect are reportedly experiencing increased delays or uncertainty thanks to Brexit (Williams, 2016), and VikingLink's developers have stated that Brexit 'does not influence the plans to build and operate' the interconnector (National Grid and Energinet, 2016).

As the UK and EU negotiate their future relationship, there are a number of both known and unknown unknowns about the outcome of what have been described as the 'most complex negotiations of all time' (Mason, 2016). Here we briefly explore the primary area of concern in this context – the UK's future engagement with the IEM – as well as other factors such as funding mechanisms and technical and regulatory cooperation.

Perhaps one of the few certainties at the time of writing is that trade in energy will continue between the UK and the rest of Europe post-Brexit, and therefore some level of engagement with the IEM is inevitable (Norton Rose Fulbright, 2016). What is uncertain at this stage is the level to which the UK will have access, or what barriers or tariffs may be introduced. Analysis has suggested that if the UK were to be excluded from the IEM, an increased cost to the UK energy system of £500 million per year would be incurred (Vivid Economics, 2016).

Several non-EU countries already have access to the IEM, setting possible precedents for future UK engagement. At one end of the scale, the UK could remain full members of the market, with minimal impact for all parties, while the other end of the scale could involve potential tariffs on trade under World Trade Organization (WTO) rules (Fieldfisher, 2017). While it is widely thought that the former would be favourable for all parties, other issues such as the involvement of the European Court of Justice and freedom of movement of people are likely to dictate the UK's future IEM relationship (Froggatt *et al.*, 2017; Kumar, 2017).

Another known unknown is the effect of Brexit on the finances of interconnector projects, in terms of access to funding, perceived investment risks, and exchange rate impacts. The UK is one of the largest recipients of funds from the European Investment Bank (EIB), and between 2012–2016, €9.3 billion was given for energy-related projects in the UK, including interconnectors (Frogatt *et al.* 2017). While the UK will not necessarily lose access to this and other EU strategic funds post-Brexit, concern has been expressed that EU-backed capital for interconnectors may no longer be available (Norton Rose Fulbright, 2016). The uncertainty that surrounds Brexit is also found to be impacting investor confidence, causing potential delays to funding decisions, with such perceived risks increasing capital costs for investors, particularly for projects scheduled to come online after 2022 (BEIS Committee, 2017; Bosch, 2017). Finally, a weakened and more volatile pound, as observed since the referendum, impacts capital costs as well as the long-term economic case of interconnectors for various stakeholders (Bosch, 2017).

If the UK were to participate in the IEM post-Brexit, it would need to ensure continued regulatory alignment in order to avoid increased complexity (Fieldfisher, 2017). The UK is a member of a number of technical institutions that define the rules of the IEM, and has historically had significant influence here. The government has signalled its intention to retain participation in these institutions, however this is uncertain, presenting the risk that the UK will go from rule-maker to rule-taker (BEIS Committee, 2017; Froggatt *et al.*, 2017; Kumar, 2017).

In May 2017, the UK's Business, Energy and Industrial Strategy Committee published findings from its inquiry into Brexit negotiation priorities for energy and climate change. Most evidence pointed out the risks that leaving the IEM and regulatory divergence would pose to energy security, costs and decarbonisation, the potential loss of influence over policy formation and impact to investor confidence (BEIS Committee, 2017). As such, the Committee recommended that in its negotiations the UK should seek continued access to the EU's financial institutions and the IEM, with no tariffs or barriers to trade, as well as continued influence of the IEM's rules through full membership of associated technical institutions (BEIS Committee, 2017). At the time of writing, the UK government is yet to respond to these recommendations.

Overall perspective and concluding remarks

As the UK moves towards a low-carbon economy, powered increasingly by intermittent renewables, it is redefining its energy relationship with its Nordic neighbours, looking away from fossil fuels and to greater electricity interconnection. Connecting the UK electricity grid to Denmark, Norway and Iceland requires high capital costs and pushes technical boundaries compared to geographically closer alternatives, but has the potential to deliver significant benefits for the UK's security of supply and consumer prices, while accelerating the decarbonisation of electricity.

Interconnection with Denmark offers benefits on both sides of the cable, despite potential increased consumer prices in Denmark, due to the synchronicity of both countries' growing wind production systems. Norway's hydro-dominated system and zero-carbon marginal generation combined with its pumped storage capacity offer unique opportunities for the UK to reduce the carbon intensity of its electricity supply and access pumped storage capacity. IceLink, the most ambitious and complex of the UK's proposed interconnectors, can deliver significant benefits to both parties, but the technical and political complexities of the cable and the required additional generation capacity needed in Iceland may generate uncertainty about the long-discussed future of this link.

All the while, the UK is in the process of leaving the EU, and attempting to negotiate an as yet unknown future relationship with the bloc. For interconnectors, while the fundamental benefits of increased interconnection remain, Brexit brings into question the extent to which the UK can be integrated into the IEM. Added to uncertainties over financing and regulatory issues, it is widely viewed that both the UK and the EU would be best served by continued barrier-free IEM access with full regulatory alignment on energy.

To incentivise interconnection links that deliver the most benefits for cost and carbon, the UK government must continue to develop its policy framework, especially addressing the current policy flaws in carbon pricing for electricity imports.

Ultimately, to meet future electricity demand with the fewest carbon emissions, and in the most cost-effective manner, the UK must continue to invest in connecting to the large renewable resources of its Nordic neighbours.

References

Aurora Energy (2016) 'Dash for Interconnection', www.auroraer.com/wp-content/uploads/2016/10/Dash-for-Interconnectors-Aurora-Energy-Research-February-2016.pdf.

BEIS (2017a) 'Renewables', in: 'Energy Trends September 2017', London, UK Government. p. 55.

BEIS (2017b) 'Clean Growth Strategy', London, UK Government.

BEIS (2018) 'Implementing the End of Unabated Coal By 2025', London, UK Government.

BEIS Committee (2017) 'Leaving the EU: Negotiation Priorities for Energy and Climate Change Policy', London, House of Commons Library.

Bjerregaard, T.R. (2017) 'Leder: Viking Link – en mørklagt forbindelse', Ingeniøren, 3 November, https://ing.dk/artikel/leder-viking-link-morklagt-forbindelse-208068.

Bosch, J. (2017) 'Interconnectors, the EU Internal Energy Market and Brexit', Imperial College London, www.imperial.ac.uk/media/imperial-college/grantham-institute/public/publications/discussion-papers/Interconnectors,-the-EU-Internal-Electricity-Market-and-Brexit.pdf.

Caird, J.S. (2017) 'Legislating for Brexit: the Great Repeal Bill'. London, House of Commons Library.

Clark, G. (2016) 'Greg Clark's Speech at the Annual Energy UK Conference', 10 November, London, UK Government.

Craig, N., Sharp, G., Baresic, D., Menezes, D. and Milhères, C (2017) 'Written Evidence from the Polar Research and Policy Initiative', Submission to the BEIS Committee Inquiry: 'Leaving the EU: Negotiation Priorities for Energy and Climate Change Policy Inquiry', London, House of Commons Library.

Danish Energy Agency (2017) 'Denmark's Energy and Climate Outlook 2017', Copenhagen, Danish Energy Agency.

Deign, J. (2017) 'Why Norway Can't Become Europe's Battery Pack', Green Tech Media, 13 March, www.greentechmedia.com/articles/read/why-norway-cant-become-europes-battery-pack.

Ea Energianalyse (2015) 'Integration af vindkraft. Viking Link og andre tiltag for integration af vind', www.ea-energianalyse.dk/reports/1550_Integration_vindkraft_viking_link_og_andre_tiltag.pdf.

Ea Energianalyse (2016) 'CO2-emission ved øget elforbrug', Copenhagen, Ea Energianalyse.

The Economist (1988) 'An Electrifying Opportunity', 9 January, London, *The Economist*.

Energinet (2017) 'Elektrisk Bro Mellem Storbritannien Og Danmark Giver Store Gevinster', https://energinet.dk/Om-nyheder/Nyheder/2017/10/27/Elektrisk-bro-mellem-Storbritannien-og-Danmark-giver-store-gevinster.

European Climate Foundation (2011) 'Power Perspectives 2030: On the Road to a Decarbonised Power Sector', The Hague, European Climate Foundation.

European Environment Agency (2016) 'Overview of Electricity Production and Use in Europe', Copenhagen, European Environment Agency.

Eurostat (2017) 'Electricity Price Statistics', http://ec.europa.eu/eurostat/statistics-explained/index.php/Electricity_price_statistics.

Fieldfisher (2017) 'Impact of Brexit for Interconnectors', www.fieldfisher.com/media/5313890/interconnectors-brexit.pdf.

Froggatt, A., Wright, G. and Lockwood, M. (2017) 'Staying Connected. Key Elements for UK-EU27 Energy Cooperation after Brexit', London, Chatham House.

Grøv, E., Bruland, A., Nilsen, B., Panthi, K. and Lu, M. (2011) 'Developing Future 20 000 MW Hydro Electric Power in Norway – Possible Concepts and Need of Resources', Trondheim, SINTEF Building and Infrastructure.

Henden, A. L., Doorman. G and Helseth A. (2016) 'Economic Analysis of Large-scale Pumped Storage Plants in Norway', *Energy Procedia*, 87: 116–123.

IHA (2017) '2017 Hydropower Status Report', www.hydropower.org/sites/default/files/publications-docs/2017%20Hydropower%20Status%20Report.pdf.

Karagiannopoulos, L. (2017) 'Iceland Magma Drilling Project May Revive Giant UK Power Cable Link', *Reuters*, 10 April, www.reuters.com/article/us-iceland-energy-britain/iceland-magma-drilling-project-may-revive-giant-uk-power-cable-link-idUSKBN17C17E.

Kirby, M. (2018) 'Map of Proposed and Existing Interconnectors', adapted from Vaughan, 2017 for this book chapter.

Kumar, K. (2017) 'Negotiating Brexit. Positive Outcomes for the UK on Energy and Climate' Green Alliance, www.green-alliance.org.uk/negotiating_brexit_energy_climate.php.

Landsvirkjun (2017) 'Submarine Cable to Europe: Overview of IceLink', www.landsvirkjun.com/researchdevelopment/research/submarinecabletoeurope.

Lodge, T. and Mahoney, D. (2017) 'The Hidden Wiring', Centre for Policy Studies, www.cps.org.uk/publications/the-hidden-wiring.

Lund, H., Mathiesen, B. V., Christensen, P. and Schmidt, J. H. (2010) 'Energy System Analysis of Marginal Electricity Supply in Consequential LCA', *The International Journal of Life Cycle Assessment*, 15(3): 260–271.

Mason, R. (2016) 'Brexit Talks May Be Most Complicated Negotiation Ever, Says Davis', *Guardian*, 12 September, www.theguardian.com/politics/2016/sep/12/brexit-talks-may-be-most-complicated-negotiation-ever-says-minister.

McGlade, C. and Ekins, P. (2015) 'The Geographical Distribution of Fossil Fuels Unused when Limiting Global Warming to 2 [deg] C', *Nature*, 517(7533): 187–190.

Met Office (2017) 'Global Circulation Patterns', www.metoffice.gov.uk/learning/learn-about-the-weather/how-weather-works/global-circulation-patterns.

Miljøstyrelsen (2017) 'Vestenvinden', http://mst.dk/friluftsliv/undervisning/naturkanon/vejr-og-himmelfaenomener/vestenvinden/.

Monforti, F., M. Gaetani and E. Vignati (2016) 'How Synchronous is Wind Energy Production among European Countries?', *Renewable and Sustainable Energy Reviews*, 59: 1622–1638.

Moore, S. (2014) 'Getting Interconnected', Policy Exchange, https://policyexchange.org.uk/wp-content/uploads/2016/09/getting-interconnected.pdf.

Mount, A., Coats, E. and Benton, D. (2016) 'Smart Investment. Valuing Flexibility in the UK Electricity Market', Green Alliance, www.green-alliance.org.uk/resources/Smart_investment.pdf.

Muñoz, I. (2015) 'Example – Marginal Electricity in Denmark', www.consequential-lca.org.

National Grid (2014) 'Getting More Connected', London, National Grid.

National Grid and Energinet (2016) 'VikingLink and Brexit', http://viking-link.com/media/1113/20160719_ngvl_viking-link-and-brexit_press-release-v10_gb.pdf.

National Infrastructure Commission (2016) 'Smart Power', www.nic.org.uk/publications/smart-power/.

Nelsen, A. (2015) 'Wind Power Generates 140% of Denmark's Electricity Demand', *Guardian*, 10 July. www.theguardian.com/environment/2015/jul/10/denmark-wind-windfarm-power-exceed-electricity-demand.

Neslen, A. (2017) 'Could British Homes Be Powered by Icelandic Volcano?', *Guardian*, 19 May, www.theguardian.com/sustainable-business/2017/may/19/british-energy-iceland-volcano-geothermal-power.

Norton Rose Fulbright (2016) 'Impact of Brexit on the Energy Sector', London, Norton Rose Fullbright.

Norwegian Ministry of Petroleum and Energy (2016) 'White Paper on Norway's Energy Policy: Power for Change', Oslo, Norwegian Government.

Norwegian Water Resources and Energy Directorate (2016) 'Electricity Disclosure 2015', Oslo, Norwegian Government.

Ofgem (2013) 'Electricity Capacity Assessment Report', www.ofgem.gov.uk/ofgem-publications/75232/electricity-capacity-assessmentreport-2013.pdf.

Ofgem (2017a) 'Cap and Floor Regime: Initial Project Assessment of the GridLink, NeuConnect and NorthConnect Interconnectors', www.ofgem.gov.uk/publications-and-updates/cap-and-floor-regime-initial-project-assessment-gridlink-neuconnect-and-northconnect-interconnectors.

Ofgem (2017b) 'Decision on the Final Project Assessment of the NSL Interconnector to Norway', www.ofgem.gov.uk/publications-and-updates/decision-final-project-assessment-nsl-interconnector-norway.

Ofgem (2017c) 'Transmission Networks: Electricity Interconnectors', www.ofgem.gov.uk/electricity/transmission-networks/electricity-interconnectors.

Pöyry (2014) 'Near-term Interconnector Cost-benefit Analysis: Independent Report. A Pöyry Report For Ofgem', London, Ofgem.

Pöyry (2016) 'Costs and Benefits of GB Interconnection', London, Ofgem.

Reed, S. (2017) 'Power Prices Go Negative in Germany, a Positive for Energy Users', *New York Times,* 25 December, www.nytimes.com/2017/12/25/business/energy-environment/germany-electricity-negative-prices.html.

Renewable Energy Foundation (2018) 'Balancing Mechanism Wind Farm Constraint Payments', www.ref.org.uk/constraints/indextotals.php.

Staffell, I., Green, R., Gross, R., Green, T., Bosch, J. and Bruce, A. (2017) 'Importing Electricity, Exporting Emissions?' in: 'Electric Insights Quarterly. Q3 2017', Drax, www.drax.com/wp-content/uploads/2017/11/ElectricInsightsReport_Q3-2017.pdf.

Strbac, G., Aunedi, M., Pudjianto, D., Djapic, P., Teng, F., Sturt, A., Jackravut, D., Sansom, R., Yufit, V. and Brandon, N. (2012) 'Strategic Assessment of the Role and Value of Energy Storage Systems in the UK Low Carbon Energy Future', London, Carbon Trust.

Vaughan, A. (2017) 'Brexit and Energy: Does "Taking Back Control" Mean Losing Power?' *Observer*, 6 May, www.theguardian.com/business/2017/may/06/brexit-energy-taking-back-control-losing-power.

Vivid Economics (2016) 'The Impact of Brexit on the UK Energy Sector', www.vivideconomics.com/publications/the-impact-of-brexit-on-the-uk-energy-sector.

Ward, A. (2017) 'Gas and Coal are Big Winners in Electricity Capacity Auction', *Financial Times*, 4 February, www.ft.com/content/1e3d6e34-ea84-11e6-893c-082c54a7f539.

Watt, H. (2017) 'Hinkley Point: The "Dreadful Deal" behind the World's Most Expensive Power Plant', *Guardian*, 21 December, www.theguardian.com/news/2017/dec/21/hinkley-point-c-dreadful-deal-behind-worlds-most-expensive-power-plant.

Williams, D. (2016) 'Iceland-UK Power Interconnector Delayed by Brexit', *Power Engineering International*, 21 October, www.powerengineeringint.com/articles/2016/10/iceland-uk-power-interconnector-delayed-by-brexit.html.

Index

Page numbers in **bold** denote tables, those in *italics* denote figures.

Printed and bound by CPI Group (UK) Ltd, Croydon, CR0 4YY

24/10/2024

01778278-0005

.